Y0-AAA-397

NONLINEAR SCIENCE THEORY AND APPLICATIONS

Numerical experiments over the last thirty years have revealed that simple nonlinear systems can have surprising and complicated behaviours. Nonlinear phenomena include waves that behave as particles, deterministic equations having irregular, unpredictable solutions, and the formation of spatial structures from an isotropic medium.

The applied mathematics of nonlinear phenomena has provided metaphors and models for a variety of physical processes: solitons have been described in biological macromolecules as well as in hydrodynamic systems; irregular activity that has been identified with chaos has been observed in continuously stirred chemical flow reactors as well as in convecting fluids; nonlinear reaction diffusion systems have been used to account for the formation of spatial patterns in homogeneous chemical systems as well as biological morphogenesis; and discrete-time and discrete-space nonlinear systems (cellular automata) provide metaphors for processes ranging from the microworld of particle physics to patterned activity in computing, neural and self-replication genetic systems.

Nonlinear Science: Theory and Applications will deal with all areas of nonlinear science – its mathematics, methods and applications in the biological, chemical, engineering and physical sciences.

Nonlinear science: theory and applications

Control and optimization J. E. Rubio

Chaos A. V. Holden (*Editor*)

Automata networks in computer science F. Fogelman Soulié, Y. Robert and M. Tchuente (*Editors*)

Simulation of wave processes in excitable media V. S. Zykov

Oscillatory evolution processes I. Gumowski

Mathematical models of chemical reactions P. Erdi and J. Tóth

Introduction to the algebraic theory of invariants of differential equations K. S. Sibirsky

Almost periodic operators and related nonlinear integrable systems V. A. Chulaevsky

Soliton theory: a survey of results A. P. Fordy (*Editor*)

Other volumes are in preparation

Fractals in the physical sciences

H. Takayasu

Manchester University Press
Manchester and New York
Distributed exclusively in the USA and Canada by St. Martin's Press

Published by Manchester University Press
Oxford Road, Manchester M13 9PL, UK
and Room 400, 175 Fifth Avenue,
New York, NY 10010, USA

Distributed exclusively in the USA and Canada by St. Martin's Press, Inc.,
175 Fifth Avenue, New York, NY 10010, USA

Reprinted in 1990

British Library cataloguing in publication data
Takayasu, H.
Fractals in the physical sciences.
1. Physics. Mathematics. Fractal sets
I. Title II. Series
530.1′5573
ISBN 0 7190 2485 4 *hardback*
0 7190 3434 5 *paperback*

Library of Congress cataloging in publication data
Takayasu, H.
Fractals in the physical sciences / H. Takayasu.
p. cm — (Nonlinear science)
Includes bibliographical references.
ISBN 0–7190–2485–4. 0–7190–3434–5
1. Science – Mathematics. 2. Fractals. I. Title. II. Series
Q158.5.735 1989
501′.51–dc20 89–14588

Typeset in Hong Kong
by Graphicraft Typesetters Limited

Printed and bound in Great Britain by
Biddles Ltd, Guildford and King's Lynn

Contents

Foreword

A line is a 1.0 dimensional object, a square is 2.0 dimensional and a cube is 3.0 dimensional. Of course, most objects in nature have contours more complicated than that of geometric figures. It should not be surprising that the dimension of a cloud, say, is not integral. Fractal theory tells us that most objects in nature have fractional dimensions; for example, 1.5 for a tree, 1.3 for a cloud.

'Fractals' were invented by Benoit B. Mandelbrot only about fifteen years ago in order to describe complicated shapes in nature. The history of fractals is short, but fractals have already been applied throughout diverse fields of science, and fractals are now indispensable in most natural sciences.

This is a textbook about fractals in natural sciences. It starts with intuitive explanations of the essence of fractals and ends up with more complicated problems.

The first three chapters are written mainly for beginners. The meaning of 'fractal' is explained without using difficult mathematics. Numerous examples of fractals in nature are introduced and computer-simulated fractals are discussed, with short program listings for personal computers.

The last three chapters are for advanced study. Theoretical models in physics are analysed and mathematical tools for fractals, such as the renormalisation group method and stable distributions, are introduced, with examples. Mathematical aspects of fractals are summarised in the final chapter. Although the topics become more technical, I am confident that the material will be accessible to anyone with a mathematical background that includes simple calculus.

This book is based on my Japanese book *Fractals* which was published by Asakura-Shoten (Tokyo) in 1986. I would like to express my sincere thanks to Shun-ichi Amari who read my Japanese book and encouraged me to write this English version. I am greatly indebted to the series editor Arun V. Holden and the editorial staff of Manchester University Press who not only corrected the English style throughout the manuscript but also suggested revisions and additions that have, I hope, made this book a better one.

I acknowledge Benoit B. Mandelbrot, Keisuke Ito and Tamas Vicsek for their encouragements and helpful suggestions. I am also grateful to Mitsuhiro Ikegami, Hisao Hayakawa, Hiroshi Kohno and Shinich Sasa for their assistance in preparing the manuscript. Ken Hattori, Takeshi Ohashi and Masaaki Kunigami kindly helped me in writing the computer programs. I would also like to thank Clarence F. Hisky, Betty Freedman, Andrew Ehrgood and Tohru Uzawa for their help on linguistic problems.

I would also like to thank all the following people for permitting me to use their figures: D. Avnir, P. Bak, V. N. Belykh, F. H. Champane, S. Lovejoy, M. Matsushita, P. Meakin, D. C. Rapaport, F. Shlesinger, R. F. Voss and T. A. Witten.

Finally, I express my gratitude to my wife Misako Takayasu for her devoted help in the publication of this book. I also thank her for stimulating discussions as a collaborator.

H. Takayasu
June 1989

1 Introduction to fractals

At first sight, fractal shapes are bewildering in their complexity. They seem too complicated to be treated analytically. However, as we get used to fractal shapes, this first impression will be proved wrong.

Chapter 1 gives an intuitive introduction to the concept of fractals for readers who are not familiar with it. One can skip almost all of this chapter since it is nearly independent of the rest of the book. The only thing which must be studied closely is the fractal dimension, the most important quantity, and one which will appear on almost every page in the following chapters.

1.1 Characteristic length

Any shape can be put into one of two categories – by whether or not it has a characteristic length. What we call 'a characteristic length' is a typical or defining length of the shape we are considering, for example, the radius of a sphere or the height of a man. We do not need to fix an exact meaning for a characteristic length, so we could take, say, the length of the legs as a characteristic length for a man instead of the height.

There exists a characteristic length for any geometric shape we learned about in elementary education – the square, the rectangular prism, the ellipse etc. Also, all artificial products, such as cars, buildings or cigarettes have their own characteristic lengths. These shapes have a property of being able to be approximated to by simple shapes with the same characteristic length. For example, seen from a distance we may regard a man as a cylinder. If one wanted to make a more man-like object, one could approximate the trunk by a rectangular prism, legs and arms by four cylinders, and the head by a sphere. If one wanted to improve still further, more and more detailed features, such as nose and fingers could be added.

Shapes with characteristic lengths have an important common peculiarity, their smoothness. The surface of a sphere is everywhere smooth, as is the side of a rectangular prism. Consider that we often regard the earth, to a first approximation, as a sphere. In this case, we have implicitly assumed that the roughness of the earth's surface is much smaller than the characteristic length of the earth, the radius; hence we do not lose any important feature by replacing the earth's surface in our mental picture by a smooth surface.

Now, let us think about the shapes belonging to the other category. It may be difficult to imagine a shape not having a characteristic length if you are not familiar with fractals. So here we start with some concrete examples. Imagine the shape of a cloud or smoke blowing up from a smoke-stack. The shape is, of course, neither a sphere nor a cylinder but is much more complicated. In order to approximate the shape we need to use a number of spheres of different sizes (if we use cubes instead of spheres the situation is the same). We cannot catch the form of a cloud by a single shape or a few elementary shapes with characteristic lengths, because complexity is the essence of its form.

Figure 1.1 shows a basic fractal shape, the Koch curve. Let us try to approximate the complicated shape by segments or by triangles. A rough approximation is shown in the top diagram of Figure 1.2, but obviously it is far from the Koch curve. The middle diagram gives the second-order approximation and the last one shows the third-order approximation. As we raise the order of approximation, the shape becomes increasingly like the Koch curve. Note that each segment in the first stage is replaced in the second by four segments which compose a shape similar to the whole shape. And the segments in the second stage are also replaced by such similar shapes in the third stage. The essence of the Koch curve is in this similarity. Actually, the Koch curve is created by repeating the above replacements to an infinitesimal scale. This kind of similarity, that the shape of a part is similar to that of the whole, is the most peculiar feature of a shape having no characteristic length and is called *self-similarity*.

Let us examine the self-similarity of the Koch curve in more detail. The part of the Koch curve in the interval $(0, \frac{1}{3})$ becomes exactly identical to the whole shape if it is magnified three times. The same can be said about the shapes in the intervals $(\frac{1}{3}, \frac{1}{2})$, $(\frac{1}{2}, \frac{2}{3})$ and $(\frac{2}{3}, 1)$. Furthermore, parts in the intervals $(0, \frac{1}{9})$, $(0, \frac{1}{27})$, ... are also similar to the whole shape. Thus in any small part there exist many miniatures of the whole with different sizes.

In the case of clouds or smoke self-similarity does not hold in the strict sense but does hold statistically. When we observe a cloud through several different magnifications we will find shapes which look very similar to the whole shape of a cloud. For any shape with characteristic length, its complexity is decreased if we observe a magnified part, but for the

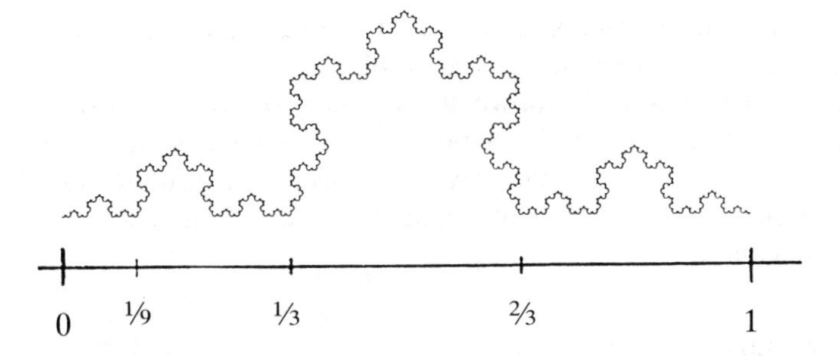

Figure 1.1 The Koch curve.

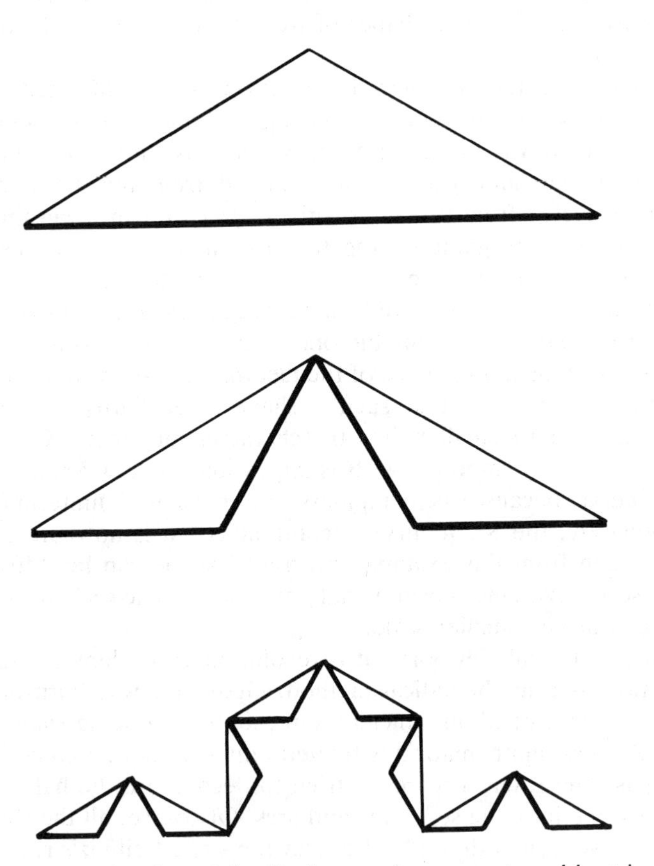

Figure 1.2 Approximation of the Koch curve by segments and by triangles.

shape of a cloud or the Koch curve the complexity does not decrease at all since any part is similar to the whole.

Self-similarity is also known as *scale-invariance*, because self-similar shapes do not change their shape under a change of observation scale. This is a very important symmetry and gives a clue to understanding complicated shapes which have no characteristic length, such as the Koch curve or clouds.

1.2 **Fractals**

We have considered a few examples of shapes having no characteristic length. There are many other examples in nature. For example, consider the shape of a coastline. When we see a coastline on a map or an aerial photograph we cannot guess the scale of the map unless there are some man-made products visible – a road or a harbour. This means that the shape of the coastline is scale-invariant, that is, it does not have any characteristic lengths. The shapes of trees or a network of blood vessels also have this quality.

The word 'fractal' is a new term introduced by Mandelbrot [1] to represent shapes or phenomena having no characteristic length. The origin of this word is the Latin adjective '*fractus*' (broken). The English words 'fractional' and 'fracture' are derived from this Latin word. As might be expected from these derivatives, '*fractus*' can mean the state of broken pieces being gathered together irregularly. We may use such a state also as a working definition for the fractal. There is no strict definition or usage for the word 'fractal' even among specialists. It is used as a generic noun, as a countable one, and sometimes as an adjective.

One interesting consequence of the fact that a shape with a characteristic length is smooth is that such a shape's peculiarity is not lost by smoothing the parts smaller than the characteristic length. On the other hand, fractals deny smoothness. It is impossible to draw a tangent line to a Koch curve because its bumpiness continues to infinitesimal scales. More precisely, the Koch curve is continuous but is nowhere differentiable. As seen from this example, no fractal shape can be differentiable since its scale-invariance ensures that any macroscopic-scale irregularity is reproduced at any smaller scale.

From a historical viewpoint, it is revolutionary to deny the use of differentiation as a mathematical method. Geometry was born in Ancient Egypt and developed in Ancient Greece by mathematicians such as Euclid. Ancient mathematicians treated only shapes that could be drawn by compass and ruler. They even tried to decompose the ballistic trajectory of a stone into line segments and arcs. Of course, all the shapes they considered were smooth. Only after Newton's and Leibniz's revolutionary work were mathematicians able to treat more complicated geometry by using differentiation. Since then, the importance of differential calculus and geometry has grown and grown. Most of modern physics is inaccessible if we cannot use differentiation. Even in Einstein's beautiful theory of gravity, space-time structure is assumed to be differentiable. The matters we shall treat here are separate from this historical trend. Fractals will lead us to a completely new geometry where the principal roles are played by complicated shapes instead of differentiable ones.

Modern physicists have made great efforts to elucidate both the smallest structures, such as atoms and elementary particles, and the largest structure, the universe. However, until recently physicists have paid little attention to structures of intermediate sizes – those comparable to our body size. This is not because they have thought intermediate-sized structures are less interesting but rather because limit structures are easier to deal with by mathematical analysis based on differentiation. In both limit sizes, there are characteristic lengths, the size of an atom or an elementary particle on the one hand, and the size of the universe on the other. In such cases, fluctuations on other scales can be smoothed out and differential geometry can be applied. In an intermediate-scale system we often miss important properties if we treat any one scale as a characteristic length and neglect others. In such a case, dominant properties result from interactions between different sizes, and the system would behave quite differently if it had only one or a few characteristic lengths. The, analytic method which takes the system to pieces does not work with complicated many-body systems. We have to cope with complexity by some other means.

The idea of a fractal is based on the lack of characteristic length or on self-similarity. Such similarities were known long ago. For example, T. Terada about 80 years ago in a scientific essay (see [2]) described cracks in metal or glass: 'Observing cone-like cracks every day through a microscope, their size seems to be getting larger and larger. Now they look as if they are mountains ... But actually they are as small as 0.1 mm.' He had found the similarity between the macro-scale world and the micro-scale world. However, he did not succeed in developing his idea any further. In order to develop the notion of self-similarity, a quantitative description is needed.

In the next section we will meet new quantities called fractal dimensions which characterise self-similarity. The key to the fractal dimension is the Hausdorff dimension which was introduced about 70 years ago as a highly mathematical concept. Mandelbrot [1] made great strides with this mathematical concept so that it can now be used to describe complicated natural shapes quantitatively.

The best way to understand the fractal intuitively is to look at computer-generated fractal shapes. The importance of computers for the study of fractals is very great – only since the recent advent of accessible, powerful computers can we visualise fractal shapes. Computers are an invaluable substitute for the analytical methods of differentiation. It could be said that the theory of fractals is cultivating an undeveloped field of science – one that has been neglected because of its complexity – with the help of the most important product of high technology, the computer.

1.3 **Fractal dimension**

Before introducing the fractal dimension, let us review the concept of 'dimension'.

We know empirically that the dimensions of a line and a plane are 1 and 2, and that we live in a 3-dimensional space. In physics we often add a time axis and consider that space-time is 4-dimensional. All of these empirical dimensions take integer values and they coincide with degrees of freedom – defined by the number of independent variables. For example, the location of a point on a line is determined by one real number; a set of two independent real numbers is needed to define a plane. If we define dimension by degrees of freedom in this way, we can consider d-dimensional space for any non-negative integer d. Actually, it is conventional in mechanics to replace the motion of m particles in 3 dimensions by the motion of one particle in $6m$-dimensional space by considering each particle's location and momentum as independent.

The dimension as defined by degrees of freedom looks very natural; however, it contains a serious flaw. Nearly one hundred years ago, Peano described a very strange curve, now called the Peano curve, which could completely cover a square. It is defined by a limit of complicated curves as shown in Figure 1.3. From this figure we see that the curve can be drawn with a single stroke and it tends to cover the plane uniformly. Since the location of a point on the Peano curve, like a point on any curve, can be characterised with one real number, we become able to describe the position of any point on a plane with only one real number. Hence the degree of freedom, or the dimension, of this plane becomes 1. This contradicts the empirical value, 2.

In order to overcome this difficulty, we have to reconsider the meaning of dimension. First we have to develop a new definition of dimension based on similarity.

Divide equally a segment, a square and a cube into similar forms of half-sizes as shown in Figure 1.4. The segment becomes 2 smaller segments, and the square and the cube become 4 and 8 smaller ones, respectively. The numbers 2, 4 and 8 can be rewritten as 2^1, 2^2 and 2^3. Note here that the exponents 1, 2 and 3 coincide with the space dimension in each case. Generalising this discussion we may say that if a shape is composed of a^D similar shapes of size $1/a$, the exponent D corresponds to the dimension. The dimension defined in this way is called the *similarity dimension*. For the Peano curve the similarity dimension becomes 2 since

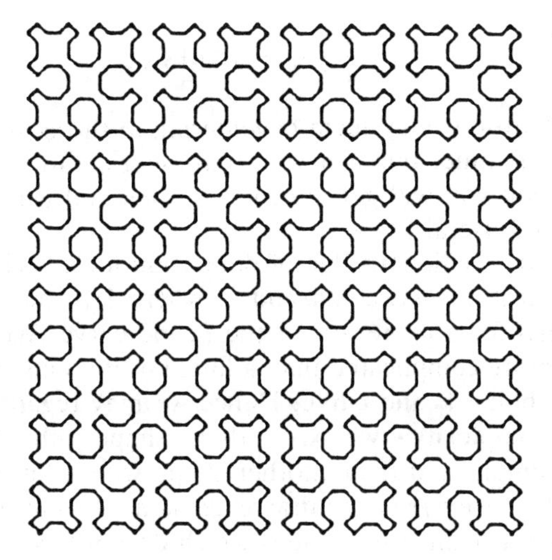

Figure 1.3 The Peano curve.

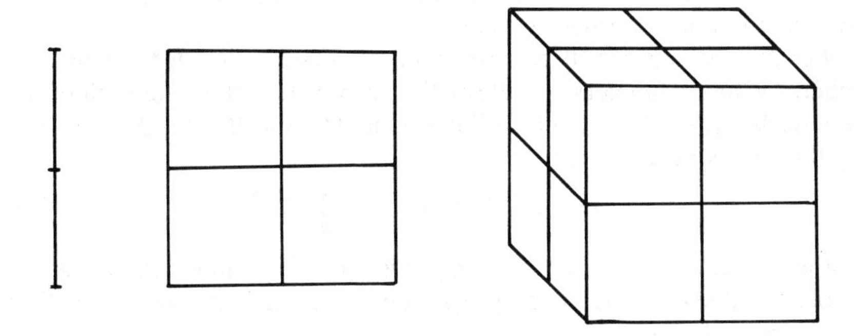

Figure 1.4 Dividing a segment, a square and a cube.

the curve is made up of 4 images of half scale (see Fig. 1.3). Using this convention, the similarity dimension does not pose any self-contradiction and it coincides with the empirical value for elementary shapes.

The most interesting feature of the similarity dimension is that its value is not restricted to integers. When a shape is composed of b similar shapes of size $1/a$, then the similarity dimension for the shape, D_s, is given by

$$D_s = \frac{\log b}{\log a}. \tag{1.1}$$

Recall the Koch curve in Figure 1.1. It is composed of four miniatures of size $\frac{1}{3}$. Hence from equation (1.1) the dimension D_s of the Koch curve becomes fractional.

$$D_s = \frac{\log 4}{\log 3} = 1.2618 \ldots \quad (1.2)$$

(Fractional dimensions may seem bewildering to those readers who meet them for the first time, this is a very natural response, but as you read through this book you will get used to the idea and will come to feel it quite natural and useful.)

What does this fractional value of dimension mean? The value for the Koch curve, 1.26, lies between 1 and 2, where 1 is the dimension for a line and 2 is that for a plane, or for the Peano curve. We know that the Koch curve is more complicated than a line, but it seems less complicated than the plane-filling Peano curve. Hence we may regard the dimension as an index of complexity – we expect that a shape with a high dimension will be more complicated than another shape with a lower dimension.

Similarity dimension is a natural extension of empirical dimension. However, it is not defined for shapes which do not have strict similarity. We need another definition of dimension which is applicable to all the shapes we want to treat. Hausdorff began the study of such a dimension in 1919. The dimension he introduced is named after him and is defined by a method of coverings as follows:

Let $D > 0$ and $\varepsilon > 0$ be real numbers. Cover a set E by countable spheres whose diameters are all smaller than ε. Denoting the radii of the spheres by $r_1, r_2, \ldots, r_k$, the D-dimensional Hausdorff measure is defined by the following equation:

$$M_D(E) \equiv \lim_{\varepsilon \to 0} \inf_{r_k < \varepsilon} \sum_k r_k^D \quad (1.3).$$

For any given set E, this quantity is proved to vary from infinity to zero at a special value of D, denoted by D_{H}. The Hausdorff dimension for the set E is defined by D_{H}.

This dimension can be defined for any shape and its uniqueness is also proved. Furthermore it can be shown that it gives the same results as the similarity dimension for the shapes we have considered above: line, square, cube, Koch curve and Peano curve. Indeed, the Hausdorff dimension is a natural generalisation of both the empirical dimension and the similarity dimension. Needless to say, it can take fractional values. Here we avoid mathematical discussions but try to understand intuitively the above definition of the Hausdorff dimension. (Some mathematical aspects of Hausdorff dimension are considered in Section 6.4.)

Let us consider covering a set by spheres; in Figure 1.5 the set is the Koch curve and the spheres are replaced by circles. These circles can be regarded as approximating the Koch curve. If the circles do not overlap much, we can estimate the total length of the approximated Koch curve simply by the sum of the radii of the circles, $r_1 + r_2 + \ldots + r_k$ ($k = 7$ in the figure). Smallness of overlapping is guaranteed by taking the infimum

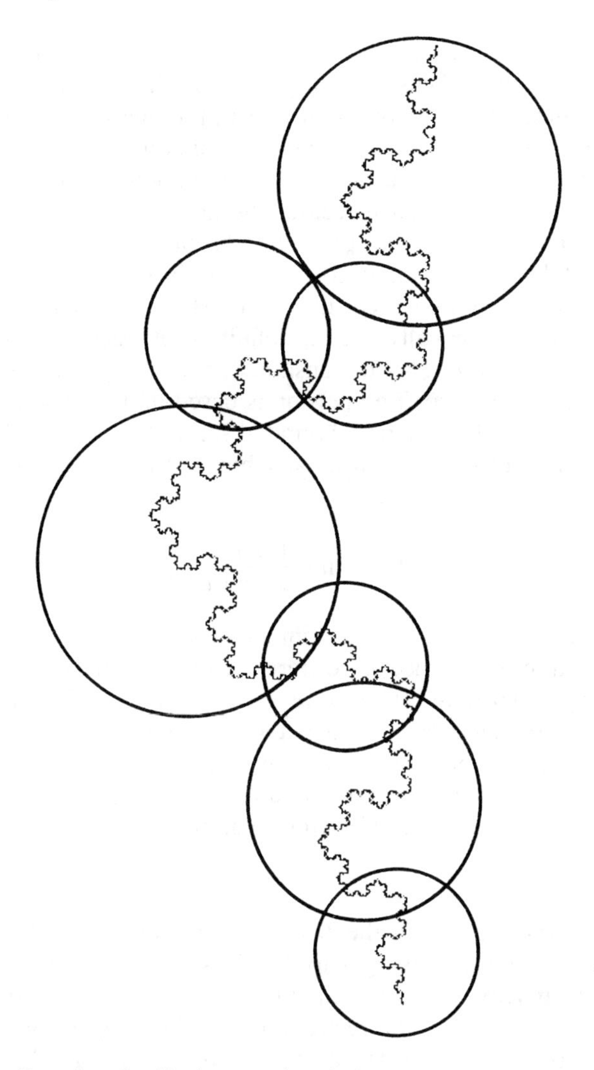

Figure 1.5 Covering the Koch curve by circles.

(lower bound) in the definition. Also, the area of the approximated Koch curve can be estimated by the summation of the circles' areas. Since the area of a circle is proportional to the square of its radius, the area of the approximated Koch curve is proportional to $r_1^2 + r_2^2 + \ldots + r_k^2$. In mathematics, the notions of length, area and volume are special cases of measures. For example, length is the natural measure of 1-dimensional objects, and area is the 2-dimensional measure. It seems natural to demand that the D-dimensional measure of a sphere of radius r be proportional to r^D. Therefore in the definition of the Hausdorff dimension we have defined the D-dimensional measure by the sum of the Dth power of each radius. Thus we can generalise the notion of measure to fractional values. By taking the limit $\varepsilon \to 0$, the measure for the approximated shape converges to that of the set's inherent value. In the case of the Koch curve, the 1-dimensional measure, namely the length, is diverging, while the 2-dimensional measure, the area, is 0. At $D = \log 4/\log 3$, the measure is mathematically proved to be finite. However, a rigorous calculation of Hausdorff dimension is generally very difficult. (See Section 6.4 for an example of a calculation of Hausdorff dimension.)

Another, more practically useful, definition of fractional dimension is the *capacity dimension*. This dimension was introduced by Kolmogorov [109] and like the Hausdorff dimension is based on the idea of coverings:

Let the considered shape be a bounded set in d-dimensional Euclidean space. Cover the set by d-dimensional spheres of identical radius $1/\varepsilon$. The capacity dimension D_c is given by

$$D_c \equiv \lim_{\varepsilon \to 0} \frac{\log N(\varepsilon)}{\log 1/\varepsilon} \tag{1.4}$$

where $N(\varepsilon)$ denotes the minimal number of spheres.

This definition may look very different from that of Hausdorff dimension, but these two dimensions have a close relationship. This time we approximate the shape by covering it with spheres of the same radius. Since we consider a most efficient covering, the D-dimensional measure of the approximated shape is estimated as $N(\varepsilon)\cdot\varepsilon^D$. For sufficiently small ε, (1.4) is equivalent to the following relation:

$$N(\varepsilon) \propto (1/\varepsilon)^{D_c}. \tag{1.5}$$

Hence in the limit of $\varepsilon \to 0$, the D-dimensional measure diverges if D is smaller than D_c and it converges to 0 if D is larger than D_c. Hence, the definition of capacity dimension may be seen as a specialised version of Hausdorff dimension with the restriction that the radii of spheres are the same. Capacity dimension often coincides with Hausdorff dimension but sometimes differs. We can prove the following inequality for any shape:

$$D_c \geq D_H. \tag{1.6}$$

These definitions of dimension are rigorous and, mathematically, there is no problem. However, a basic difficulty occurs when applying them in physics. In both definitions dimensions are defined in the limit $\varepsilon \to 0$, but length 0 is an unphysical concept in consequence of the uncertainty principle. In addition to this 'inner-length scale' below which scale invariance breaks down, there is in each physical problem also an 'outer-length scale' above which scale invariance does not hold. Thus, experimentally, there must be limitations on the observed scaling range on both sides. Hence we have to improve the definition of dimension further in order to make it applicable to complicated shapes in nature.

Many physically feasible methods of defining dimension have been devised recently. They are classified into five categories as follows:

(1) changing coarse-graining level;
(2) using the fractal measure relations;
(3) using the correlation function;
(4) using the distribution function;
(5) using the power spectrum.

Before explaining these methods, two points should be noted. One is the limitation of scale range. Any shapes in nature must have upper and lower limits. Fractal properties can be observable only between these limits. For example, we have mentioned that the shapes of clouds are statistically self-similar; however, the similarity breaks down if we observe it at the scale length of the earth or at a microscopic length. These limits often play important roles in physical problems, for example, they restrain physical quantities from diverging. However, we will not touch on this problem in the following discussion in order to keep it simple. (Readers who are interested in this problem are recommended to read Section 6.1, where a scale-dependent dimension is introduced.)

The other remark is on uniqueness. There is a problem as to whether dimensions obtained by different methods agree with each other. This is a basic but difficult problem – we cannot give a general answer. In some cases they agree but in other cases they differ. Hence, to be rigorous, we have to distinguish each dimension by a different name. However, if we restrict the subject to shapes in nature the difference is known to be small compared with observation error. Therefore we are going to name all these dimensions simply as fractal dimensions. We use fractal dimension as a generic term for dimensions that can take fractional values.

Changing coarse-graining level (called box-counting methods)

Here, we will again begin by approximating fractal shapes by some basic shapes having characteristic lengths such as circles, line segments, squares and cubes. For example, we will try to approximate a complicated curve like a coastline, shown in Figure 1.6, by a set of line segments of length r. To approximate the curve by segments, the following procedure is applied. First, let one edge of the curve be a starting point, and from there draw a circle of radius r. Then connect the starting point and the point where the circle intersects the curve by a straight line. We regard this intersection as the new starting point and repeat the above procedure until it comes to the other edge. We denote by $N(r)$ the number of segments that are needed to approximate the coastline. As we change the

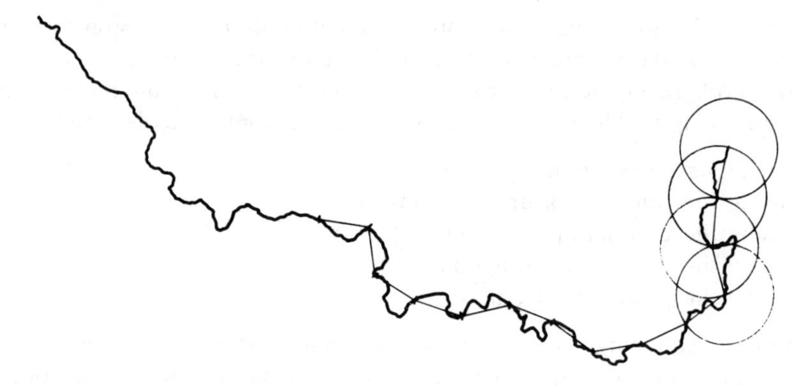

Figure 1.6 Approximating a coastline by line segments.

unit length of the segments r, the total number of segments $N(r)$ will of course vary. If the coastline is a straight line, it satisfies the following relation:

$$N(r) \propto \frac{1}{r} = r^{-1}, \tag{1.7}$$

but we cannot expect this relation to hold for complicated coastlines. As we make the length of segments r smaller, we will see more tiny structures that are not noticed when r is large, so we need more segments to cover the complicated coastline than a straight coast. Let us take the Koch curve to show this. We can see from Figure 1.2 that the Koch curve satisfies the relation of $N(\frac{1}{3}) = 4$, $N((\frac{1}{3})^2) = 4^2$, ... Thus, since $(\frac{1}{3})^{-\log_3 4} = 4$, we have

$$N(r) \propto r^{-\log_3 4} \tag{1.8}$$

The exponent $\log_3 4$ appearing here is identical to the similarity dimension or Hausdorff dimension of the Koch curve. The exponent 1 of r in (1.7) also agrees with the dimension of a straight line. Thus in general, for a certain curve, if it has the relation

$$N(r) \propto r^{-D}, \tag{1.9}$$

we may call D the *fractal dimension* of the curve. As we shall see later, fractal dimensions of actual coastlines or trajectories of random walks are often measured in this manner.

There is another method which has wider applicability and is more convenient for calculations by computer. In this method, we first divide the space into cubes with edge length r, then count the number of cells needed to cover a given shape. As an example, we measure the fractal dimension of a set of points on a plane. First, divide the plane into squares with side length r. Then we count the number of squares which

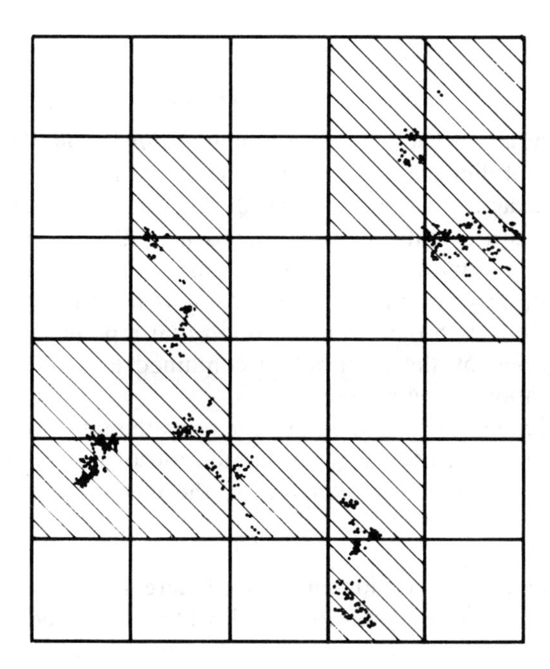

Figure 1.7 Coarse-graining by squares.

contain at least one point, and denote this number by $N'(r)$ (see Fig. 1.7). If $N'(r)$ satisfies the relation

$$N'(r) \propto r^{-D}, \tag{1.9'}$$

when the length of r is changed, we can say this distribution of points is D-dimensional. For the case where D is an integer, as it is for a straight line or a plane, it is quite definite that D fits the empirical dimension. This method is not only applicable to distributed points, but also to the analysis of more complicated figures such as rivers with many tributaries.

The information dimension, is another extension of the dimension defined above. It applies especially to stochastic distribution of points. We divide a space into cells of edges length r, as before, and calculate the probability $P_i(r)$ that one randomly chosen point is in the ith cell. This is a kind of coarse-grained observation on a scale of length r. Now the information $I(r)$ is defined by

$$I(r) \equiv -\sum_i P_i(r) \cdot \log P_i(r), \tag{1.10}$$

where

$$\sum_i P_i(r) = 1.$$

If $I(r)$ varies as

$$I(r) = I(0) - D_{\mathrm{I}} \log r, \tag{1.11}$$

when changing r, we call D_{I} the *information dimension*. The meaning of this definition of dimension may not be clear, but it is easy to show that it corresponds exactly to the empirical dimension when it has an integer value. If points are uniformly distributed in d-dimensional space, $P_i(r)$ is simply proportional to the size of the cells, so it is given by $P_i(r) \propto r^d$. By substituting this into (1.10) we get $D_{\mathrm{I}} = d$ from (1.11).

Also, it agrees with the intuitive notion that a point in D-dimensional space is specified by the information contained in D coordinates, each consisting of about $-\log r$ digits.

Both the dimension D defined by (1.9′) and the information dimension D_{I} give the dimension for a given distribution of points and it is easily proved that these two dimensions satisfy the inequality

$$D \geq D_{\mathrm{I}}. \tag{1.12}$$

The information dimension introduced here is a special case of the generalised information dimension which will be discussed in Section 6.1.

Using the fractal measure relations

In this method we define the dimension by using the fact that fractal shapes have measures of non-integral dimension.

As we multiply the side length of a cube by 2, the surface area, being the 2-dimensional measure, becomes 2^2 times as large as that of the original cube, and the volume, being the measure of 3 dimensions, becomes 2^3 times as large as the original volume. In general, for any non-fractal object, the following relation holds between its length L, area S and volume V:

$$L \propto S^{1/2} \propto V^{1/3}. \tag{1.13}$$

This relation means that $S^{1/2}$ and $V^{1/3}$ become k times as large if L is magnified k times. If there is any quantity, X that increases by 2^D when we change the size by a factor of 2, we may say the quantity is D-dimensional. This quantity X then satisfies the relation

$$L \propto S^{1/2} \propto V^{1/3} \propto X^{1/D}. \tag{1.13′}$$

Consider the Koch curve again as the example. In this case, the quantity that has a non-integral dimension is the length of the curve. In fact, by magnifying the Koch curve by 3, the length of the curve becomes $4(=3^{\log_3 4})$ times as long as the original. So, this curve's length has a dimension of $\log_3 4$.

There are several ways to determine the dimension from this relation.

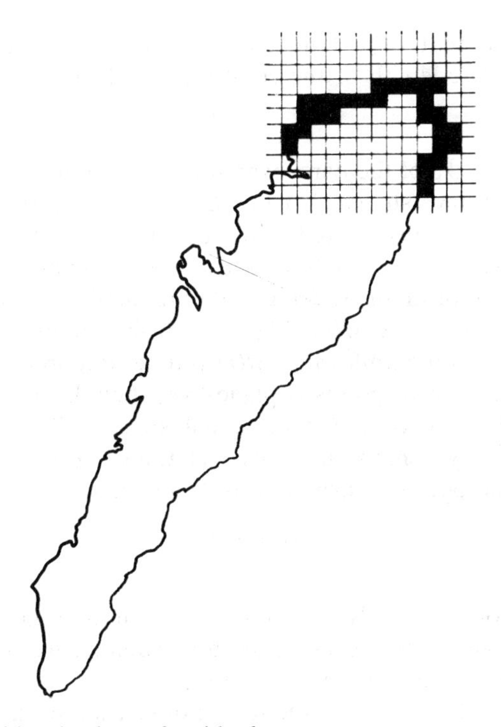

Figure 1.8 Digitising the shape of an island.

Think of measuring the fractal dimension of the coastline of islands. Let S be the area of an island and X be the length of the coastline. First, we draw a Cartesian lattice on the underlying plane and make it as fine as practically possible. Then we paint black the squares which include the coastline (Fig. 1.8), and count the number of black squares (denoted by X_N) and the number of white ones surrounded by the black squares (S_N). Of course, these numbers are approximately proportional to the length and the area, respectively – $S \propto S_N$ and $X \propto S_N$. By repeating this procedure for many islands of different sizes, we obtain a set (S_N, X_N). If we can find the relation

$$S_N^{1/2} \propto X_N^{1/D} \tag{1.14}$$

then we will agree that the fractal dimension of the coast is D. The difference between this method and the method of coarse-graining described previously is that now we do not change the size of the unit squares, but keep it as small as possible. Instead, by looking at a multitude of islands, we compare islands of different size.

We can approach the same problem without using the relation between the area and length of coastlines, but instead, use the relation between length of a coastline and the linear distance L between the end-points. As we look at one part of a long coastline, we investigate the relation between L and the length of the coastline X_N. If we find the relation

$$L \propto X_N^{1/D} \tag{1.15}$$

as we change the size of the parts examined for various combinations of L and X_N, we will have shown that this D is the fractal dimension.

For a set of points distributed in space, for instance the distribution of stars in the universe, we can define fractal dimension in a similar manner. Think of a sphere of radius r. We denote the number of points which are included in the sphere by $M(r)$ (Fig. 1.9). If the points fall on a straight line and are distributed uniformly, $M(r)$ will be proportional to r; $M(r) \propto r^1$. If the distribution of points is plane-like, then $M(r) \propto r^2$. For points distributed uniformly over 3-dimensional space, $M(r)$ should be proportional to r^3. By generalising these relations, we may say the fractal dimension of the points' distribution is D and that

$$M(r) \propto r^D \tag{1.16}$$

is satisfied.

$M(r)$ need not necessarily be a number of points. When we try to find the mass distribution of the universe, for example, we may denote the total mass in a sphere of radius r by $M(r)$.

When we actually try to find the fractal dimension for a set of points, a problem occurs as to where to position the centres of the spheres. Averaging over all data obtained by changing the position of centre will give a good result, but it is usually very time-consuming. Also, different results

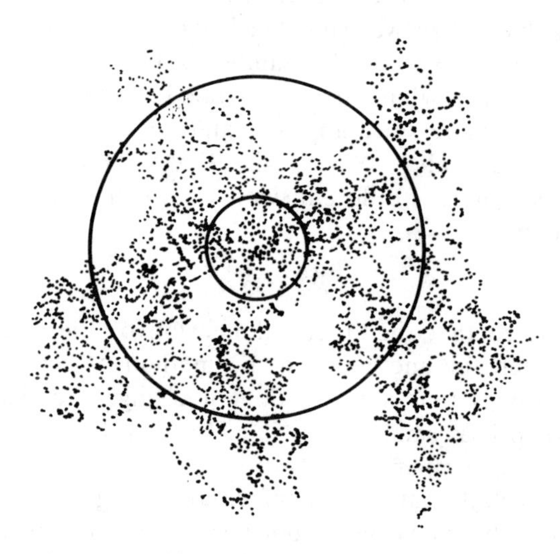

Figure 1.9 Counting the number of points within a sphere of radius r.

can be obtained from different (geometric, arithmetic, ...) averages, This will be discussed in Chapter 6 in detail. A simpler way is to centre the sphere at the centre of mass of the distribution. When the distribution of the points is uniform and fractal, the relation given by (1.16) may be obtained by this method. (However, spatial fluctuation in the value of D has recently become an important theme of research – see Section 6.1.)

Using the correlation function

The correlation function is a fundamental statistical quantity, and we can find the fractal dimension by using it.

Let $\rho(x)$ be the density at a position x of a certain quantity that is randomly distributed through space. Then the correlation function $c(r)$ is defined as

$$c(r) \equiv \langle \rho(x)\, \rho(x + r) \rangle. \tag{1.17}$$

Here $\langle . \rangle$ denotes an average. The average here can be either over an ensemble at fixed x or over the point x. If the distribution is uniform and isotropic, the correlation function is a function only of the distance r between the two points.

In theoretical models, we often assume that the exponential function $\exp(-r/r_0)$ or the Gaussian $\exp(-r^2/r_0^2)$ is the correlation function. But these functions do not possess the fractal property because both of them have a characteristic length r_0. Any pair of points whose distance is less than r_0 are strongly related to each other, but the correlation decays rapidly for $r \geq r_0$.

On the other hand, when the distribution is fractal, the correlation function follows a power law (see Section 6.4.2). Then there is no characteristic length and the rate of decrease of the correlation is always at the same rate. For instance, if the correlation is something like

$$c(r) \propto r^{-\alpha}, \tag{1.18}$$

then the correlation becomes $2^{-\alpha}$ times smaller as the distance of the two points becomes 2 times longer. The relation between the exponent α and the fractal dimension D is simply[1]

$$D = d - \alpha \tag{1.19}$$

where d denotes the dimension of the space. To check this equation, we consider the integrated mass within radius r, $M(r)$, introduced in the preceding method. From the definition of $c(r)$, it follows that $M(r)$ is calculated as

$$M(r) = \int_{|s|<r} \mathrm{d}s\; c(s)/P(0) \propto \int_0^r S^{d-1}\, S^{-\alpha}\, \mathrm{d}s\; \alpha\; r^{d-\alpha} \tag{1.20}$$

Comparing this with (1.16) we have (1.19).

If the correlation function $c(r)$ scales as in (1.18) its Fourier transform $F(k)$, which is called the power spectrum, also follows a power law. Indeed, when $0 < d - D < 1$,

$$F(k) = 4 \int_0^\infty \mathrm{d}r \cos(2\pi kr)\ c(r) \propto k^{d-D-1}. \tag{1.21}$$

Using this relation, we can estimate the fractal dimension from the power spectrum. As is well known we can directly obtain $F(k)$ by a wave-scattering experiment, and hence the fractal dimension can be estimated from light- or neutron-scattering data.[2]

Using the distribution function

Many sizes of crater appear in pictures of the Moon's surface, but we cannot know how big or small a crater is just by looking at one picture. We cannot tell whether the diameter of the crater is 100 km or 50 cm since both look very much the same. This feature is symptomatic of the lack of a characteristic length of a crater. We can deduce the fractal dimension from the distribution function.

Let $P(r)$ be the probability that an arbitrarily chosen crater has radius greater than r. This probability is related to the probability density $p(r)$ by

$$P(r) = \int_r^\infty p(s)\ \mathrm{d}s. \tag{1.22}$$

A change of scale corresponds to transforming r to $\lambda \cdot r$. The above-mentioned fractal property of the distribution requires the invariance

$$P(r) \propto P(\lambda \cdot r), \tag{1.23}$$

for any positive λ. The only functional form that satisfies (1.23) is the power law:

$$P(r) \propto r^{-D}. \tag{1.24}$$

The exponent D appearing here gives the fractal dimension of the distribution in the following sense. Consider the situation when we observe the craters with resolution r, that is the craters smaller than r are invisible. If we count the number of observable craters, the number should be proportional to $P(r)$. By changing the resolution from r to $2r$, the number of observable craters will decreased by 2^D. Hence, the exponent D in (1.24) corresponds to the fractal dimension D in (1.9′) which is defined by coarse-graining. However, it is not recommended that we call this D a dimension if r does not denote a one-dimensional measure, that is, a length. In that case, we should simply regard D as a parameter characterising the distribution.

The power law distribution (1.24) has profound connections with the stable distribution discussed in Section 5.2.

Zipf's law, well-known in the field of social science, is very closely related to fractals. A famous application of Zipf's law is found in the population of cities. If the cities of a country are ranked in the order of their population, then the product of each city's rank and population is roughly constant. That is, the population of a city is inversely proportional to its rank. This law holds in many other fields such as the amount of imports for countries or the frequency of particular words in works of literature (see Section 2.5.5).

We can show that Zipf's law is a special case of the fractal distribution of (1.24). Assume a distribution of size is given by that equation. Then the size of the considered object, r, and its rank k will approximately satisfy the relation

$$r \propto k^{-1/D}, \tag{1.25}$$

because k is proportional to $P(r)$. (Note that there are k objects whose sizes are larger than r, if the rank of the object of size r is k.)

Using the power spectrum

When we investigate the statistical features of a spatial or temporal random process, we can quite often easily get the power spectrum $S(f)$ by simply transforming the variation into electric signals and passing it through appropriate filters. We can judge from the spectrum whether a given fluctuation is fractal or not. Furthermore, if it is a fractal we can calculate its fractal dimension.

As for the spectrum, changing the level of coarse-graining of observation corresponds to a change of the cutoff frequency f_c. Here the cutoff frequency is the upper limit frequency of observation, that is, components of frequencies higher than f_c are neglected. Since fractals are invariant under a change of observation scale, the spectrum of a fractal signal must be invariant under the change of the cutoff frequency. The only spectrum $S(f)$ with this property is again the power law:

$$S(f) \propto f^{-\beta}. \tag{1.26}$$

When a spectrum is given by this power law, an important relation exists between the exponent β and the fractal dimension D of the graph of the signal:[3]

$$\beta = 5 - 2D, \tag{1.27}$$

where $1 < D < 2$. Flicker, or $1/f$ noise, observed in many electric circuits, has $\beta \doteqdot 1$, so its fractal dimension is roughly 2. If we plot a graph of voltage of $1/f$ noise as a function of time, the fluctuation will almost uniformly cover the 2-dimensional plane.

This method can be extended to higher-dimensional spaces. For example, we can obtain the fractal dimension of the relief of landscape or solid surfaces. For the case of landscape, denoting by $S(f)$ the power spectrum of fluctuation of altitude along a line, the fractal dimension of the surface, D $(2 < D < 3)$, satisfies the following relation if $S(f)$ follows the power law of (1.26):

$$\beta = 7 - 2D. \tag{1.28}$$

Notice that (1.28) can be obtained from (1.27) by replacing D by $D - 1$. This simple relation between the fractal dimension of a surface and that of its cross-section comes from general properties of fractal sets (see Section 6.4).

Notice the difference between the spectrum $S(f)$ discussed here and the spectrum $F(k)$ in (1.21) appearing in the correlation function method. It may seem confusing that we have different fractal dimensions from the power spectra $F(k)$ and $S(f)$. We use the same symbol D for both fractal dimensions, but we should notice that it has very different meanings. For $F(k)$, D means the dimension of a spatial distribution, while for $S(f)$ it is the dimension of a curve or a surface. We should use them properly, depending on our purposes.

1.4 Some basic fractals

1.4.1 *The Cantor set and the devil's staircase*

The Cantor set is one of the most popular fractals and often appears in any introduction to fractals. It has very many applications. Its definition is as follows:

Trisect a line segment $[0, 1]$ and erase the centre part $[\frac{1}{3}, \frac{2}{3}]$. Then trisect again the remaining segments $[0, \frac{1}{3}]$ and $[\frac{2}{3}, 1]$, and erase each centre part $[\frac{1}{9}, \frac{2}{9}]$ and $[\frac{7}{9}, \frac{8}{9}]$, respectively. Trisect further the remained segments and erase the centre parts. Continuing this process to infinity, we get the Cantor set (Fig. 1.10).

In other words, the Cantor set is the set of points in the interval $[0, 1]$ which are expressed only by 0 and 2 when they are written in ternary representation. Mathematically speaking, this set is complete and nowhere dense, that is, any limit point of this set belongs to this set and in any interval in $[0, 1]$ there always exists at least one sub-interval that does not contain any part of the set. It may be obvious that the fractal dimension D of this set is given as

Figure 1.10 The Cantor set.

$$D = \frac{\log 2}{\log 3} = 0.6309 \ldots \tag{1.29}$$

In Section 6.4 we will show that this value is exactly equal to Hausdorff dimension.

Suppose some quantity is uniformly distributed in the interval [0, 1] with density 1. If it is cut and contracted to the interval $[0, \frac{1}{3}]$ and $[\frac{2}{3}, 1]$, the density will be 3/2 for both intervals $[0, \frac{1}{3}]$ and $[\frac{2}{3}, 1]$, and 0 for the interval $[\frac{1}{3}, \frac{2}{3}]$, provided the total amount is invariant. If each of the parts is cut and contracted again to the intervals $[0, \frac{1}{9}]$, $[\frac{2}{9}, \frac{1}{3}]$, $[\frac{2}{3}, \frac{7}{9}]$, $[\frac{8}{9}, 1]$, the density on these intervals will be $(3/2)^2$ and elsewhere it will be 0. Letting $C(x)$ be the density in the limit where this contracting process is repeated infinitely often, $C(x)$ will be infinite on the Cantor set and otherwise zero.

The function $d(x)$, called the 'devil's staircase' is the function defined by the integral of $C(x)$:

$$d(x) = \int_0^x C(s)\, \mathrm{d}s, \tag{1.30}$$

and is shown in Figure 1.11. From this figure, we find that the derivative of this function is zero almost everywhere. It is a strange step-like function, as suggested by its name. Strange functions like this example are used in theoretical physics (Section 4.3) and observed in experiments (Section 2.4).

1.4.2 *The Sierpinski gasket*

The Sierpinski gasket is a generalised version of the Cantor set in 2-dimensional space. It is obtained by continuously taking out the centres of triangles on a plane (see Fig. 1.12). As we can deduce from the self-similarity of this figure, its fractal dimension is

$$D = \frac{\log 3}{\log 2} = 1.585 \ldots \tag{1.31}$$

Extended versions, for example, a fractal gasket in 3-dimensional space, can easily be created in the same way.

1.4.3 *De Wijs's fractal (also called the binomial multi-fractal)*

When the mineralogist de Wijs was studying the distribution of minerals in rocks he found it to be well approximated by the following model:

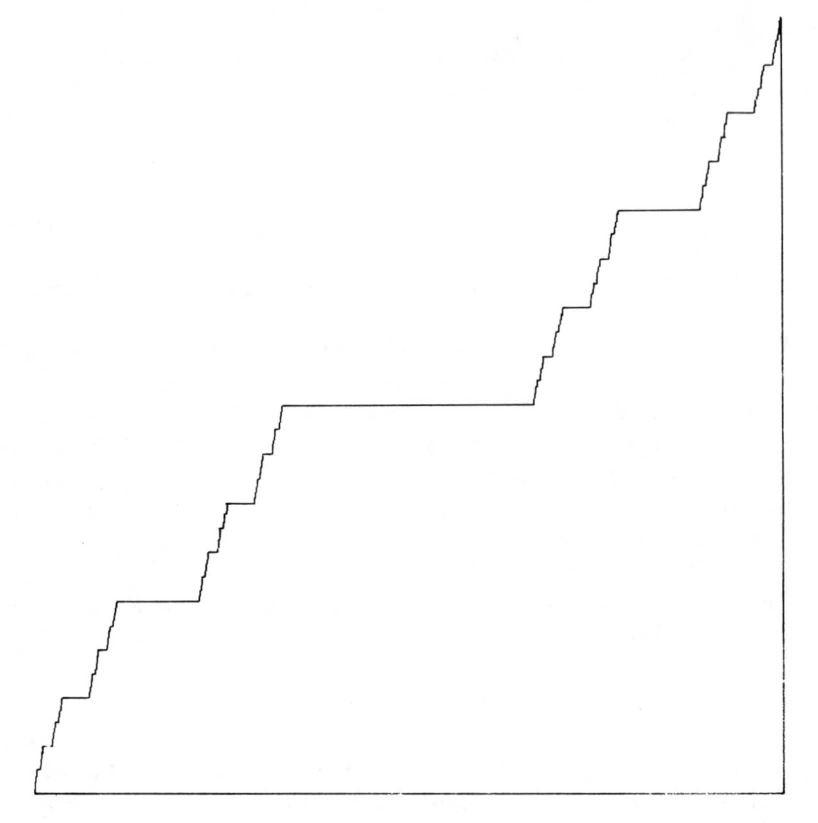

Figure 1.11 The devil's staircase.

Figure 1.12 The Sierpinski gasket.

Suppose a rock contains a certain kind of mineral whose total mass is M. If this rock is bisected, one piece of the rock contains αM mineral and the other $(1 - \alpha)M$. If each of these rocks is bisected further, then the four pieces of rocks with equal volume contain $\alpha^2 M$, $\alpha(1 - \alpha)M$, $(1 - \alpha)\alpha M$ and $(1 - \alpha)^2 M$ of the mineral respectively. That is, in every bisection, the mineral is divided in the ratio $\alpha : 1-\alpha$.

When the ratio α is 1/2, the mineral distributes uniformly. However, it can be shown experimentally that α is not 1/2; it takes an almost constant value independent of the step of division (Fig. 1.13). We call the limit of such distribution de Wijs's fractal. Its fractal dimension is given by

$$D = -\{\alpha \log_2 \alpha + (1 - \alpha) \log_2 (1 - \alpha)\}. \tag{1.32}$$

For $\alpha = 1/2$, D becomes 1, otherwise $D < 1$ and in fact for $\alpha = 1$ or 0, $D = 0$. Actually, this dimension is the information dimension defined by (1.11). This fractal is an example of a non-uniform fractal and is fully characterised by an infinite number of dimensions (see Section 6.1).

De Wijs's fractal is different from the Cantor set in that any integral over a finite interval takes a positive value if $D > 0$. This indicates that the capacity dimension is always 1. As D gets smaller, the distribution becomes more localised. When D is very close to 0, almost all the mineral seems to be concentrated at one point. If we break such rock into pieces, every piece will contain a finite quantity of the mineral, but most of the total mineral is in one piece. Although the amount could be very small, any rock on the Earth might contain any mineral, gold, for example. But most gold is localised in particular countries. And in these countries, rich deposits are restricted to some small areas, and in these areas even richer places are localised in smaller areas. De Wijs's fractal is a model of such a

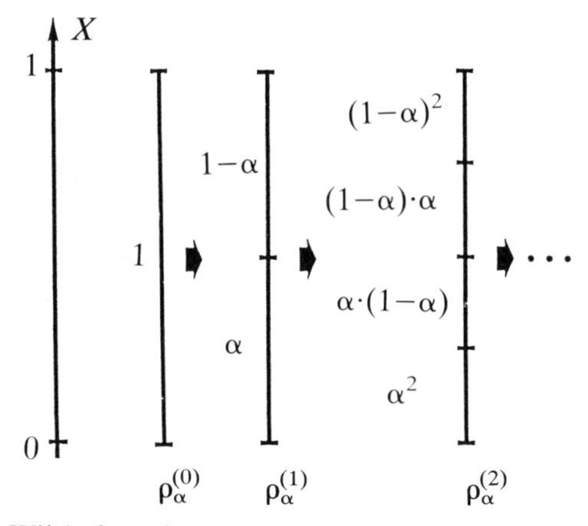

Figure 1.13 De Wijs's fractal.

distribution and the fractal dimension quantitatively describes the self-similarity.

Denote de Wijs's fractal of Figure 1.13 by $c(x)$. What shape has the integral of this function? Figure 1.14 gives the answer. This is a very strange function. It is continuous and almost everywhere differentiable with vanishing derivative. It seems paradoxical but this function is really increasing in any finite interval. Intuitively, we think of the derivative as an indication of the rate of variation of a function, so we are apt to think that a function must be constant if its derivative is almost everywhere zero. However, such intuition is denied by this strange function which is called *Lebesgue's singular function*.

Here is an interesting metaphor of this function. Let the abscissa in Figure 1.14 be a man's age, where he was born at $x = 0$ and dies at $x = 1$. The ordinate is the age that he feels. In day-to-day life he feels no change in his body (the derivative is 0), although he is aware of the change of his body when he looks at his photograph taken some years ago

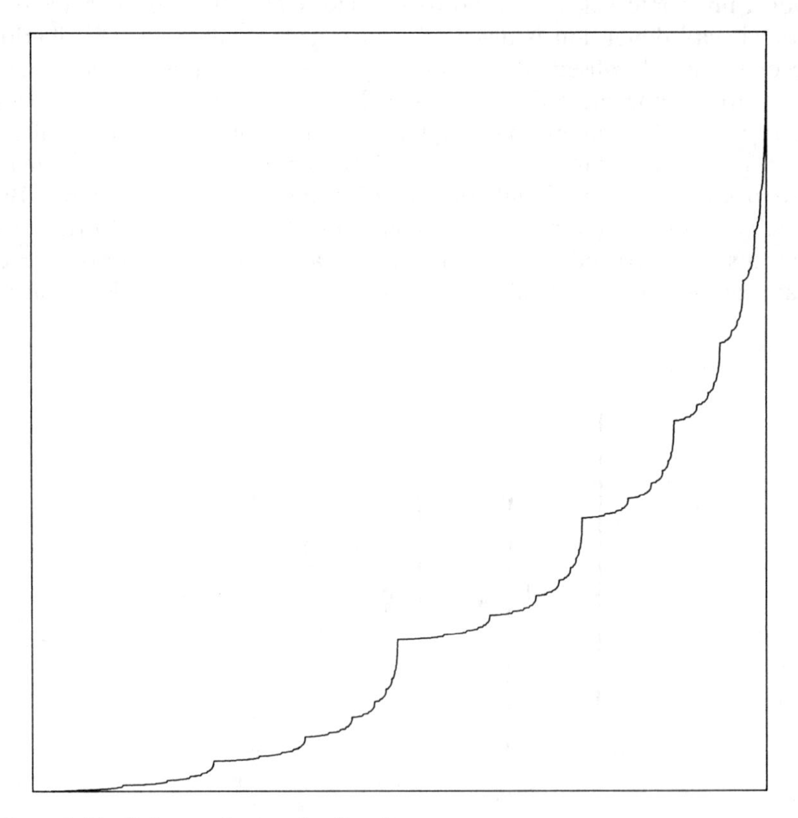

Figure 1.14 Lebesgue's singular function.

(increase in finite interval). As he ages he faces the weakening of his body (the change is more obvious near $x = 1$). Time mercilessly flows, no matter how young he thinks he still is!

1.4.4 *Lévy dust*

We have examined regular fractals, but of course there are random fractal models too. Lévy dust is a random distribution of points designed as a model of the distribution of stars in the universe. Figure 1.15 is an example of Lévy dust drawn by a computer. It might remind us of a spiral nebula. It is drawn by the following procedure:

Consider a random walk whose distribution of jump lengths follows a power law, as in (1.24). The direction at each jump is chosen completely at random. Plot a point at each place the walker jumps to. The resulting set of points is called Lévy dust or Lévy flight. (In Section 3.7 there is an example of a suitable program.)

The points seem distributed to form clusters of many sizes. The fractal dimension of this set is given by the exponent D in (1.24).

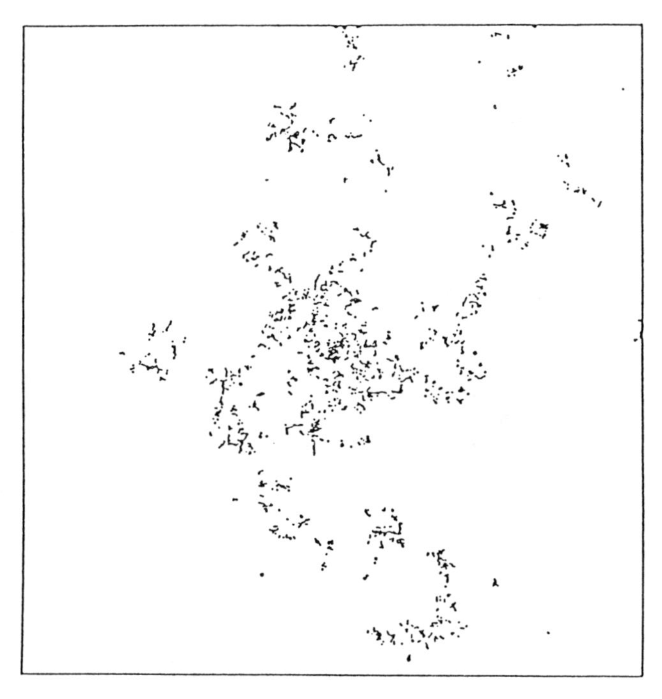

Figure 1.15 Lévy dust.

Notes

1 The fractal dimension defined by the correlation function is denoted by D_2 or ν_2 in Chapter 6, and is distinguished from D_c, D_I and D_H.

2 Denoting the differential cross-section of scattered wave by $\sigma(k)$, we have $\sigma(k) \propto k^{-D}$ from (1.21). This relation is valid for volume scattering. When surface scattering is dominant the formula is known to be modified as $\sigma(k) \propto k^{D-6}$ for a fractal surface of dimension D. See [74] for derviation.

3 This relation is exact for fractional Brownian motion, see Sections 3.7.3, 5.4 and 6.4.7.

2 Fractals in nature

In this chapter we are going to study some of the surprising number of fractals that occur in nature. They will be considered under the headings of earth science, biology, space science, physics and chemistry, and 'other topics' which includes social sciences. The aim of this chapter is to introduce the use of fractals in real problems.

2.1 Earth science

2.1.1 *Landscape*

It has already been mentioned that the shape of coastlines is fractal. Here, we consider some actual data. In Figure 2.1 the lengths of some coasts in Japan as measured on various scales are plotted on log-log graph paper. The experimental data points are approximately on a straight line in each case. The length of a coast is given by the product of the length r of unit scale, and the number $N(r)$ of unit lengths. The gradient a of this graph and the fractal dimension D of the coast are related by

$$a = 1 - D. \tag{2.1}$$

The value of D is not universal but is within the range $1 < D < 1.4$ with a mean value of 1.2. The more deeply indented a coastline, the larger the value of D.

It might be thought that for a non-fractal curve the log-log plot of r and $N(r)$ would also be linear. Look at the lowest graph in Figure 2.1, which is the plot for a circle of radius 8 km. It clearly bends when the length scale approaches the radius. As mentioned in the preceding chapter, the log-log plot of r and $N(r)$ for a shape with a characteristic length never becomes linear.

The relief of the Earth, composed as it is of mountains and valleys, is also characterised by a fractal dimension. Since a coastline is actually a

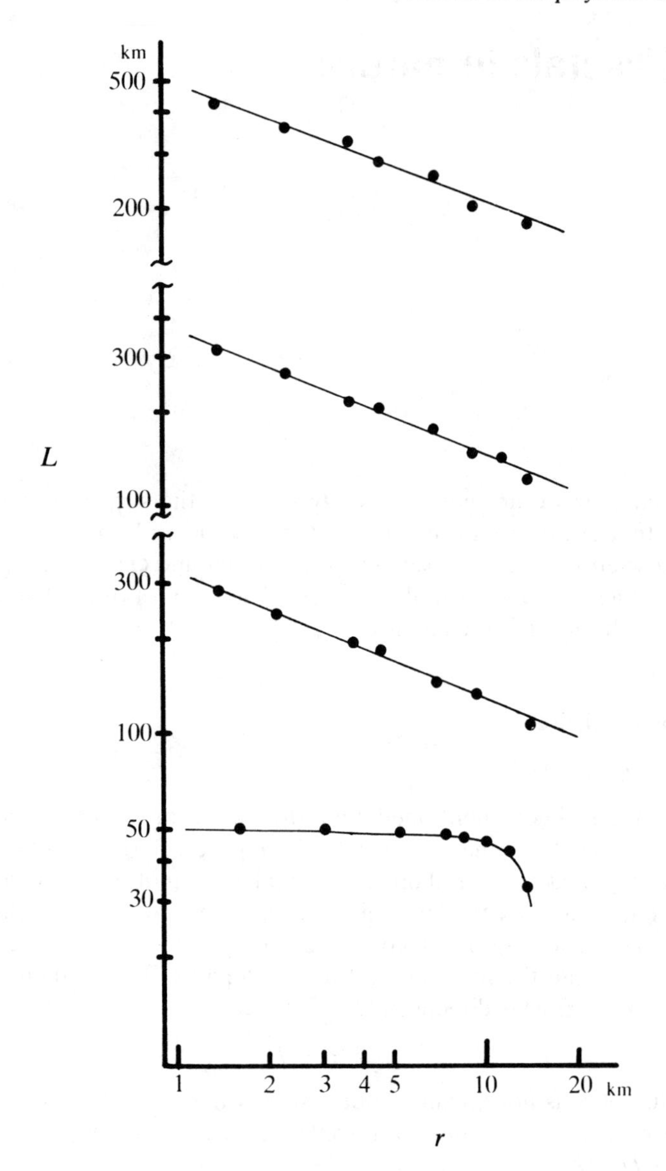

Figure 2.1 Coarse-grained total length v. unit length. The upper three are for actual coasts; the lowest is for a circle of radius 8 km.

contour line of the Earth's relief, the fractal dimension of that relief is given approximately by $D + 1$, where D is the fractal dimension of coastlines. Hence, the dimension of the Earth's relief is about 2.2. (See the discussion of the intersection of fractal sets in Section 6.4.)

The power spectrum of vertical fluctuation of the road as observed in a moving car [3] is known to be proportional to $f^{-2.5}$ where f is frequency in Hz. Combining this result with (1.28) gives a value of 2.25 for the fractal dimension of the Earth's relief, coinciding nicely with the estimate in the previous paragraph.

2.1.2 *Rivers*

Rivers have several fractal properties. Consider first the shape of the mainstream. In the field of potamology the best-known empirical law is that known as *Hack's law* [4]. This asserts that the following relation holds between the length L (km) of the mainstream and the area A (km^2) of the drainage basin:

$$L \propto 1.89\, A^{0.6}. \tag{2.2}$$

This can be rewritten in the form of (1.14):

$$A^{1/2} \propto L^{1/1.2}, \tag{2.3}$$

and hence the fractal dimension of the mainstream can be seen to be 1.2. For all rivers, the fractal dimension of the mainstream calculated individually by the coarse-graining method falls in the range 1.1 to 1.3, with a mean value of 1.2.

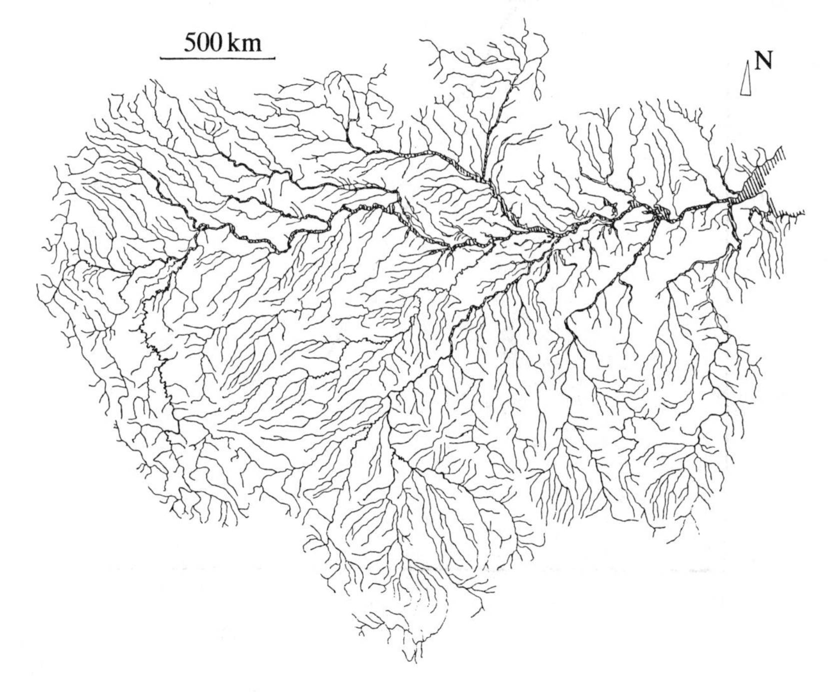

Figure 2.2 The Amazon.

The shape of a whole river system is also fractal. Using a map like the one shown in Figure 2.2, we can evaluate the fractal dimension by the box-counting method. In Figure 2.3, the number of covering boxes, N, is plotted against box size using a log-log scale. The points are clearly on a straight line whose gradient gives the fractal dimension as 1.85 for the Amazon. The fractal dimension of the Nile has been obtained in the same way as 1.4. This result suggests that the dimension of a river is larger in a region of higher rainfall.

If rain were to fall continuously on a region, then what shape river would be formed? If the system is to remain stationary, then rainfall would have to be carried away immediately. Thus, the river must cover the whole region. Hence, in such a limiting situation the fractal dimension of the 'river' would be equal to 2.

Another fractal property of rivers can be found in the temporal change of water level. The temporal fractal dimension of the Nile, estimated

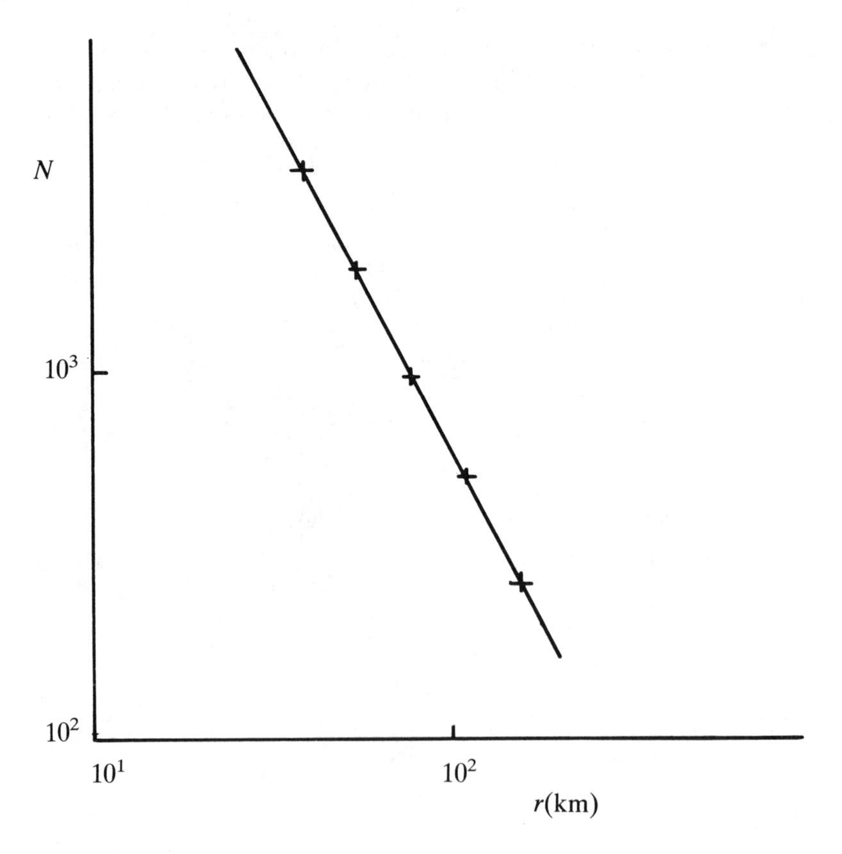

Figure 2.3 Number of squares v. coarse-graining length for Figure 2.2

from the data of annual variation of lowest water level, is about 1.1 [1]. Since water level is governed by rainfall, the change of weather is also likely to be fractal.

2.1.3 *Earthquakes*

Earthquakes are among the most interesting fractal phenomena since they have so many different fractal properties.

It is well known that the distribution of earthquake magnitude follows the *Gutenberg–Richtèr law*:

$$\log N(M) \propto -b \cdot M, \quad b \doteqdot 1, \tag{2.4}$$

where M denotes the magnitude and $N(M)$ is the number of earthquakes whose magnitude is greater than M. In the neighbourhood of Japan, earthquakes with magnitude greater than 6 occur approximately 7 times a year, while the number with magnitude greater than 5 is to about 70. Extrapolating the Gutenberg–Richter law to $M = 1$, earthquakes of magnitude greater than 1 should occur on average every minute.

Magnitude is approximately proportional to two-thirds of the logarithm of released strain energy; hence (2.4) can be rewritten as the following power law:

$$N(X) \propto X^{-2b/3}, \tag{2.5}$$

where X denotes the energy of an earthquake. This relation shows the first fractal property of earthquakes. Note that the exponent $2b/3$ cannot be called the fractal dimension since energy is not a quantity measured by length.

A second fractal property or scale invariance can be found in the spatial distribution of earthquakes. The distribution of epicentres is not uniform, for they are localised in clusters. The errors in estimating fractal dimensions are high, but the fractal dimension of the worldwide distribution of earthquakes has been quoted as being in the range 1.2 to 1.6 [5].

The third fractal property of earthquakes is found in the frequency distribution of aftershocks. The numerical density of aftershocks n(t) decays following a power law:

$$n(t) \propto t^{-P}, \quad P \doteqdot 1. \tag{2.6}$$

This relation, called *Omori's formula*, expresses a temporal fractal property called the long time tail (see Sections 2.4.9 and 4.2.2).

A fourth property is that the power spectrum of a seismic wave often obeys an inverse power law in some frequency range.

2.2 Biology

2.2.1 *Lungs and blood vessels*

The area of a surface in a finite region becomes infinite if the surface has a fractal structure on an infinitesimal scale. Lungs are organs that utilise this characteristic. By repeating self-similar ramifications, the lung's surface makes a fractal with dimension 2.2 [1]. Of course, this fractal property is not continued to an infinite scale but the surface area becomes much larger than that of a smooth ball with the same volume. Since oxygen and carbon dioxide have to be exchanged at the lung surface at each breath, the efficiency of the lung is much enhanced by this fractal property.

Blood vessels also have fractal properties. Since blood has to be

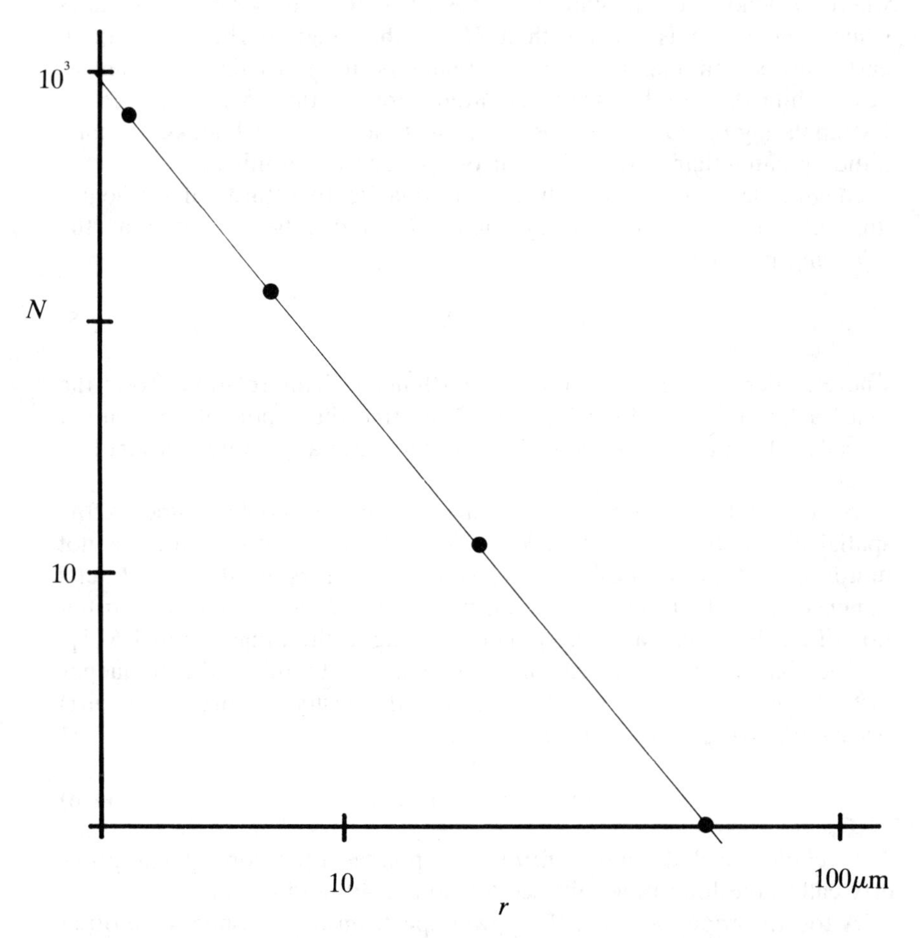

Figure 2.4 Diameter distribution of blood vessels in a bat's wing.

carried to every cell in the 3-dimensional body, the capacity dimension of blood vessels should be 3.

The diameter distribution of blood vessels in a bat's wing, ranging from capillaries to arteries, is shown in Figure 2.4 [6]. The points plotted on a log-log scale lie on a straight line following the power law:[1]

$$N(r) \propto r^{-D}, \quad D \doteqdot 2.3, \tag{2.7}$$

where r denotes the diameter and $N(r)$ gives the number of vessels thicker than r. The exponent 2.3 can be regarded as the fractal dimension of the size distribution of blood vessels.

An interesting property relates to bifurcations of blood vessels. Let the diameter of the vessel before the bifurcation be r, and the diameters after bifurcation be r_1 and r_2. Then the new exponent Δ is defined by the following relation:

$$r^{\Delta} = r_1^{\Delta} + r_2^{\Delta}. \tag{2.8}$$

This exponent Δ is called the *diameter exponent* and is known to be constant over a wide range of bifurcations [1]. The value of Δ is 2.7 [7]. Since Δ is larger than 2, the diameters satisfy the inequality

$$r^2 < r_1^2 + r_2^2. \tag{2.9}$$

This means that the vessel's cross-section is larger after a bifurcation, and therefore the velocity of blood flow decreases.

The diameter exponent has been obtained for many other branching structures. For example, $\Delta = 2.0$ for trees and for river widths. For a neuronal tree and for bronchi the exponents are 1.5 and 3.0, respectively [1].

2.2.2 *Tree shapes and insect numbers*

The branching of a tree resembles that of a river, and so tree shapes are expected to be fractal. This is confirmed by analysing photographs of trees by the box-counting method. Estimated fractal dimensions of trees [8] range from 1.3 to 1.8, with a mean value of 1.5. We may say that trees are fractals with dimension of about 1.5.

Some readers might object that this value is too small because they believe that branches of a tree spread fully in 3-dimensional space. This intuition is not correct. If the dimension of the tree were larger than 2, then a photograph of a tree would be densely covered by branches according to the projection properties of a fractal set (see Section 6.4.6). However, trees can be seen through – the fractal dimension must be less than 2.

The surface roughness of plants is closely related to the distribution of insect sizes. Small insects can live in smaller gaps than large insects. For insects, the surface area of a tree is not fixed but depends on their body sizes. Therefore the fractal dimension of surface roughness is expected to govern the ecosystem of insects.

Let us assume that the fractal dimension of a plant's surface is 2.5. An insect of size 0.1 cm can use $10^{2.5-2} = 3.16$ times larger area than an insect of 1 cm. It is empirically known that the metabolic rate is proportional to the 0.75th power of the weight. Hence the 1 cm insect consumes approximately $1000^{0.75} = 178$ times more energy than the 0.1 cm insect. Combining these effects, we may predict that the number of insects of size 0.1 cm is approximately $3.16 \times 178 \doteqdot 560$ times greater than the number of 1 cm insects. The observed size distribution of plant-dwelling insects follows a power law which roughly corresponds to this ratio [8].

2.3 Space science

2.3.1 *Distribution of mass*

Stars are not uniformly distributed in the universe – they form galaxies and the galaxies form clusters. Mass in the universe seems to have a tendency to cluster.

A cluster of galaxies may contain from one hundred to several thousand galaxies. A typical diameter would be about 20 million light years. It is known that clusters of galaxy tend to form super-clusters; and there are some regions, which are called voids, of size several hundred million light years where no galaxy exists.

The correlation function for galaxy distribution is found to follow a power law. The fractal dimension of mass distribution estimated from this power law is about 1.2 [1, 9]. This value is much smaller than 3, the dimension of the space. No theory has yet succeeded in explaining this value.

2.3.2 *Diameter distribution of craters and asteroids*

In Chapter 1 it was mentioned that the distribution of crater diameters is fractal [10]. Here we consider an example of actual data. In Figure 2.5, the cumulative number of craters on a region of the Moon is plotted against crater diameter r on a log-log scale; the points are clearly on a straight line. The distribution can be written as

$$N(r) \propto r^{-D}, \quad D \doteqdot 2.0, \tag{2.10}$$

where $N(r)$ denotes the number of craters whose diameters are larger than r. This D may be regarded as the fractal dimension of crater size distribution. The value $D = 2.0$ seems to be universal, for it is also valid for craters on Mars and Venus.

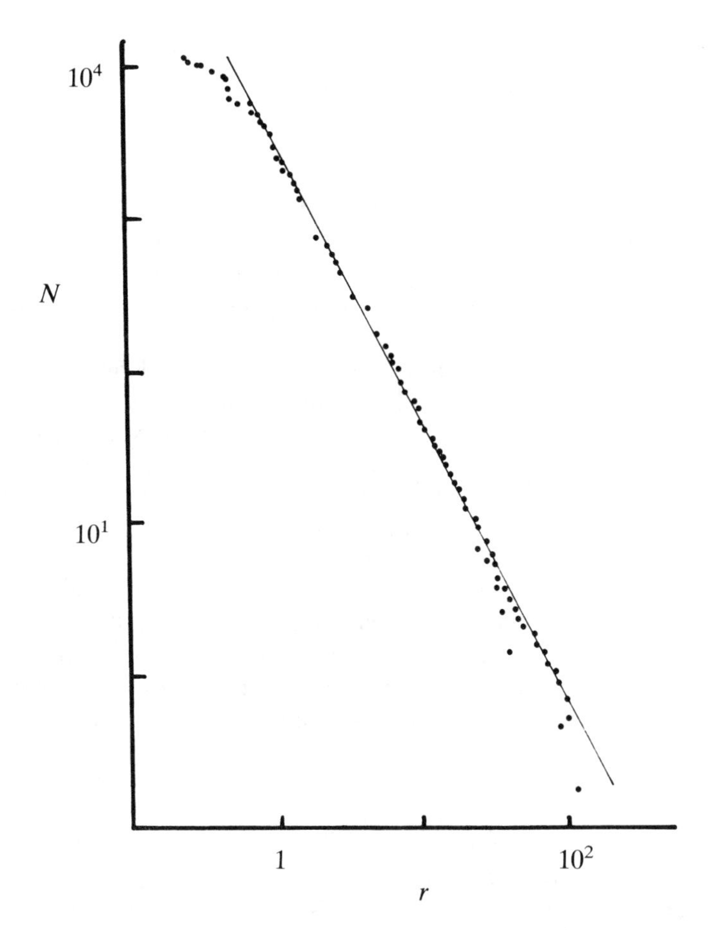

Figure 2.5 Diameter distribution of craters on the Moon [10].

Considering that craters are made by meteorites, we might expect that the size distribution of meteorites is also fractal. Actually the distribution of meteorite mass follows a power law with $D = 2.3$ for meteorites larger than 100 kg. Most smaller meteorites are burnt up by friction with the atmosphere and those arriving on Earth fail to follow the power law.

The size distribution of asteroids is also known to be governed by a fractal distribution of estimated dimension $D = 2.1$.

We may be able to develop a unified theory to explain these results. In the study of brittle fracture the distribution of splinters is known to follow a power law. For example, when a rock is shattered with a gun, the distribution of splinter size is a power law with D nearly equal to 2. Thus if asteroids and meteorites are broken pieces of a larger body, it would be a natural consequence for their size distribution to follow a power law with D nearly equal to 2. The theory of fracture has not been sufficiently developed for more precise statements.

2.3.3 *The rings of Saturn*

One of the remarkable observations made by the Voyager 2 probe was of the extremely fine structure of the Saturn ring system. The rings can be observed from the Earth using only a pair of binoculars, but even with an astronomical telescope each of the rings appears as a homogeneous band. These bands were known to be composed of small particles, but it seemed that they were segregated into bands with clearly defined edges. The Voyager 1 and 2 provided startling images that the rings themselves are composed of thousands of thinner ringlets each of which has a clear boundary separating it from its neighbours.

This structure of rings built up of finer rings has some of the properties of a Cantor set. The classical Cantor set is constructed by taking a line one unit long, and erasing its central third. This process is repeated on the remaining line segments, until only a banded line of points remains – see Figure 1.10. Although the mechanism of formation of the Saturn ring system is unclear, one theoretical approach shows that gravitational interactions between Saturn and its satellites, and interactions between the ring particles naturally lead to a structure with the properties of a Cantor set [11].

2.3.4 *The interplanetary magnetic field*

As Voyager 2 approached Saturn, it made high-resolution measurements of the interplanetary magnetic field. It was found that the magnetic field fluctuates self-similarly over time scales from 20 s to 3×10^5 s [12]. The fractal dimension of the graph of the field is estimated to be 5/3, and the corresponding power spectrum becomes $f^{-5/3}$ from Equations (1.26) and (1.27). Since the solar wind carries the magnetic field pattern past the observer, the observed time series should show the spatial pattern of the magnetic field. The Kolmogorov spectrum for homogeneous, isotropic, stationary turbulence is also $f^{-5/3}$, as will be explained in Section 5.3. The Voyager 2 measurements are consistent with the observation of an inertial range of turbulence.

2.4 **Physics and chemistry**

2.4.1 *The surface of solids*

At the molecular size range, the surfaces of many solids are fractals [13], that is, surface geometric irregularities are characteristically self-similar at different resolutions. This has been confirmed by the experimental measurement of adsorption rates. Let us consider spherical molecules of radius r being adsorbed; $N(r)$ is the number of adsorbed moles. Since molecules cover the surface with a single layer, $N(r)$ corresponds to the number of coverings in Equation (1.9). Here the molecules play the role of the measuring stick!

In Figure 2.6, adsorbed moles are plotted against particle cross-section ($\propto r^2$) for silica gel. The points are clearly on a straight line in log-log scale. The fractal dimension of the surface is obtained from the gradient of the line, and turns out to be 2.97 ± 0.02, which indicates that the surface of a silica gel particle is very complicated and is nearly volme-filling.

The fractal dimensions of solid surfaces are known not to be constant, but to range from 2 (flat) to 3 (volume-filling). They depend not only on the composition of the material but also on how it was produced.

This fractal property has a deep influence on the efficiency of chemical reactions. Roughly speaking, the higher the dimension, the greater the efficiency, since most chemical reactions take place on the surface. One

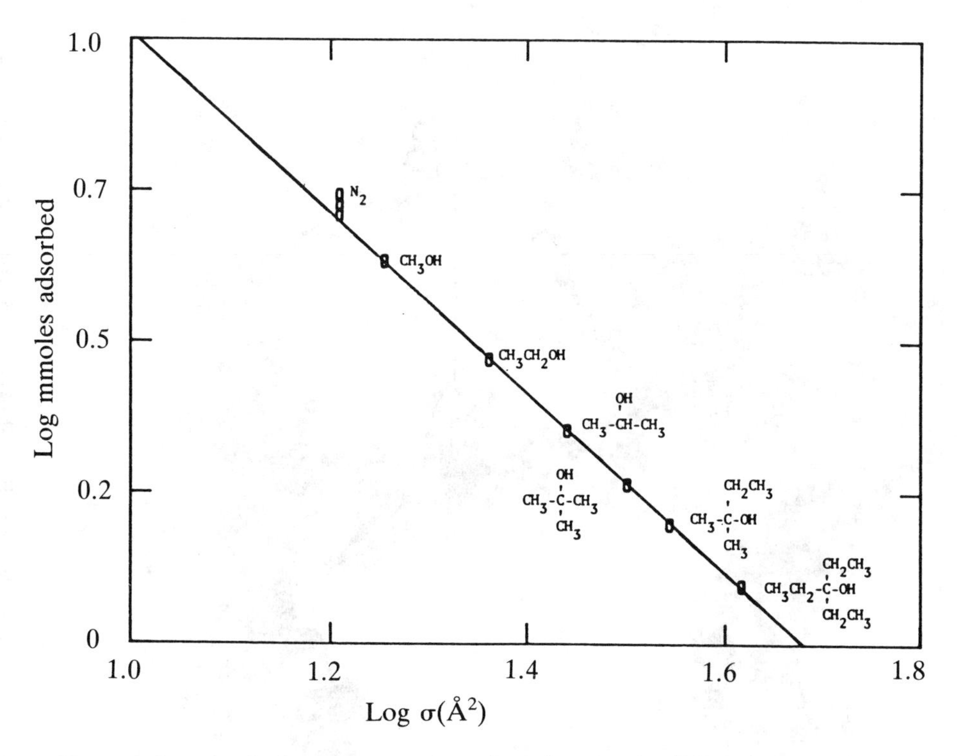

Figure 2.6 Adsorbed moles v. cross-section of molecules [13].

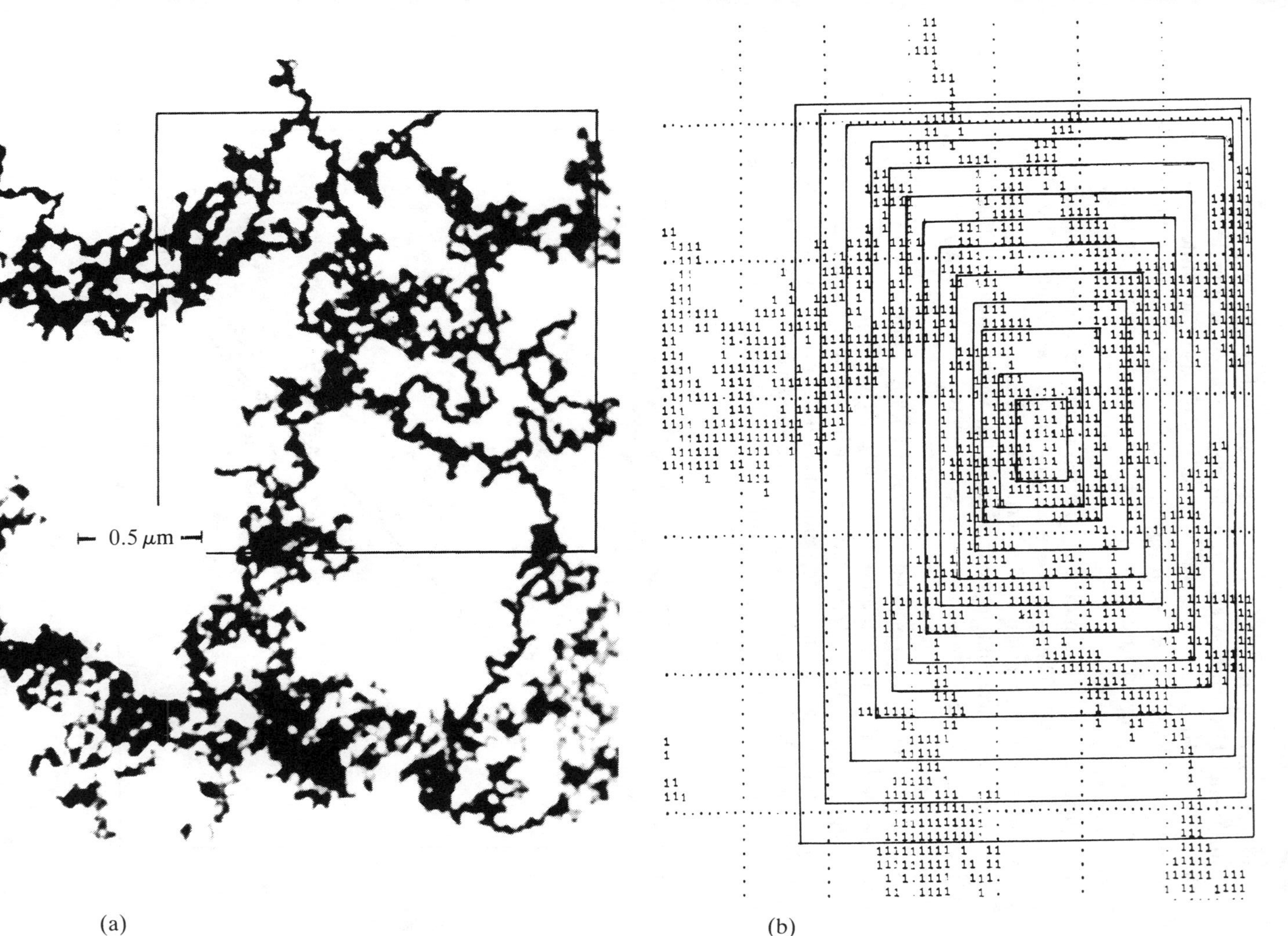

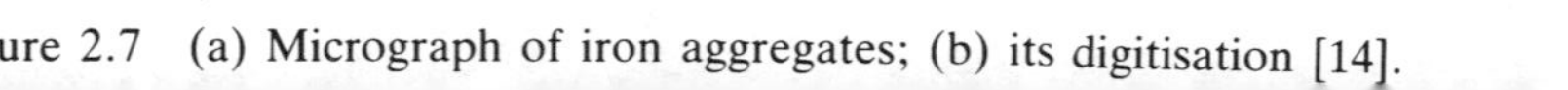

Figure 2.7 (a) Micrograph of iron aggregates; (b) its digitisation [14].

reason silica gel is used as a drying reagent is because it has a high surface-to-volume ratio.

2.4.2 *Aggregation*

An aggregate of fine particles, such as those of soot, also has a fractal structure. Figure 2.7(a) is a photograph of aggregated particles taken by an electron microscope [14]. The size of the particles is about 35×10^{-10} m, a which is smaller than the resolution of the microscope. Digitising the photograph as in Figure 2.7b enables the fractal dimension to be estimated as 1.5 from Equation (1.16) and 1.6 from Equation (1.20). These values are largely independent of the aggregated material.

When a pair of electrodes is put into a solution of a metallic ion, the metal is deposited on one of the electrodes. Under some conditions the deposited metal forms a shape which resembles a tree or leaf. Figure 2.8 shows an example of a 'metal leaf' made by the following simple experiment [15]. Prepare zinc sulphate (concentration about 2 M) in a Petri dish about 5 mm deep, and pour n-butyl acetate over it in order to make an interface. Put a negative electrode at the centre of the dish, and surround it by a positive pole made of zinc as shown in Figure 2.9. Application of a potential difference of about 5 volts between the electrodes will cause a growth of metal leaf in a few minutes. This metal leaf is a typical fractal and its dimension is estimated as about 1.7.

The fractal dimension of aggregates in 3-dimensional space has also been investigated [16], and a dimension obtained of 2.5–2.6.

These aggregations have been of great interest to physicists and many interesting results have been obtained [17]. Computer simulations have indicated that such complicated structures can be obtained by simply

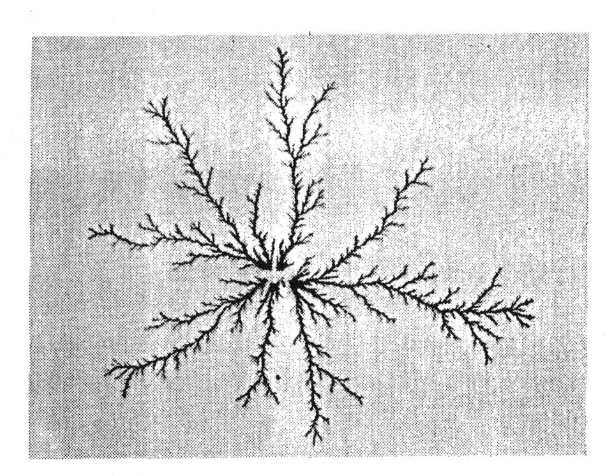

Figure 2.8 Dendrites [15].

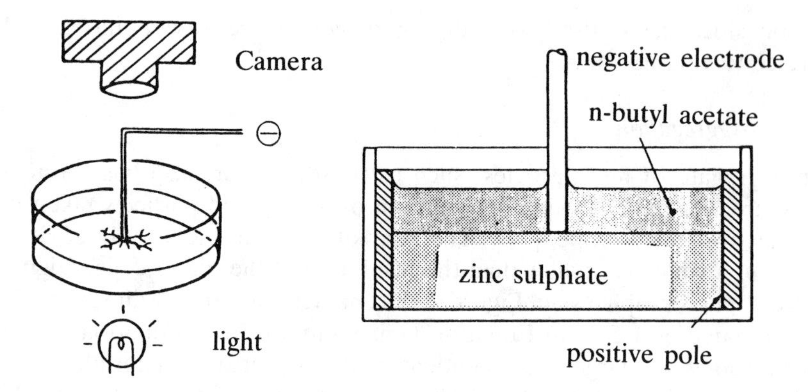

Figure 2.9 The apparatus for making dendrites.

assuming thermal random walks and irreversible adhesion of fine particles (see Section 3.1).

2.4.3 *Viscous fingering*

When fluid with low viscosity is injected under pressure into highly viscous fluid, the injected fluid forms a dendritic structure resembling the aggregation in Figure 2.8. Such structures are called *viscous fingers* [18].

This phenomenon occurs in oil extraction. In drilling for oil, five holes are usually drilled; four at the corners of a square and one at the centre. If water is injected into the centre hole, oil is extracted from the corner holes, for the water drives the oil through those holes. The efficiency of extraction is decreased if water spreads with a fractal structure since most oil is then left behind between fractal branches.

When making experiments on viscous fingers, we should pay attention to the following. Generally speaking, fractal structures do not appear where stabilising forces are stronger than destabilising ones. Therefore we should decrease surface tension and diffusion across the inferface in order to enhance the fractal property. When water is used as the low-viscosity fluid, one of the best high-viscosity fluids is polysaccharide polymer solution. This is a hydrophilic non-Newtonian fluid and its viscosity is 1000 to 10 000 times greater than that of water. Diffusion becomes negligibly small if the rate of growth is not too slow. We can observe viscous fingers by injecting water into viscous fluid sandwiched between glass plates. It is reported that the fractal dimension of a viscous finger takes values from 1.4 to 1.7 [19].

2.4.4 *Discharge patterns*

The shape of a lightning discharge also resembles that of rivers. The fractal property of the discharge pattern has been analysed and the fractal dimension is estimated to be about 1.7 [20].

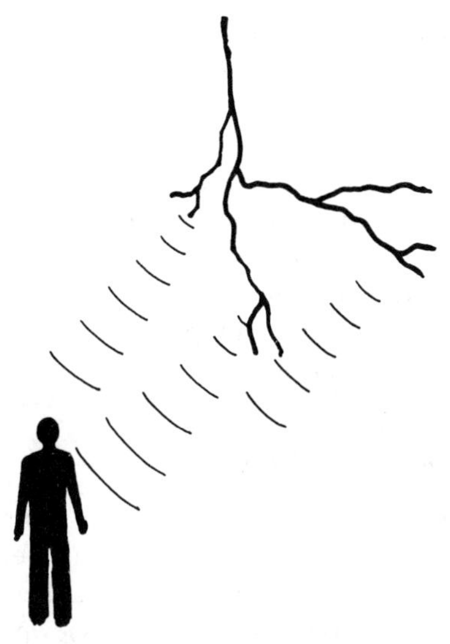

Figure 2.10 Sounds from thunder.

Aggregations, viscous fingers and discharge patterns have very similar shapes and their fractal dimensions are nearly equal to each other. Such similarity is not an accidental result; and it has recently been demonstrated that these phenomena satisfy a common equation (see Section 3.4).

In connection with discharge patterns we may also consider the relation to the resulting sound, the roll of thunder. The sound created by a small discharge is a pulse of shock wave; the discharge pattern of lightning is the source of pulse-like sounds [21]. What sound would we hear if sound sources are distributed fractally? Sounds from near sources are immediate and strong, and those from far sources are delayed and weakened. Superposition of these sounds becomes the roll of thunder (see Fig. 2.10).

2.4.5 *Polymers*

The complicated 3-dimensional configuration of polymers has been clarified by X-ray analysis: the long molecules in solution are not linear but bent randomly like tangled threads. Fractal analyses have been carried out, measuring total mass within a radius r, and the fractal dimension has been obtained for various polymers [22]. The value of the dimension is about 1.6 for thread-like polymers. As will be discussed in Section 3.5, this value corresponds to the 5/3 fractal dimension of a self-avoiding random walk.

Fractal structures of polymers affect some physical properties such as Raman scattering. It is known that the temperature-dependence of Raman relaxation is given by T^{3+2D} where T is the absolute temperature and

D is the fractal dimension [22]. Raman scattering is governed by the spectrum of proper vibration in which molecules are regarded as a set of coupled oscillators. This shows that the proper vibration spectrum depends on the fractal dimension.

It should be noted that a thread-like polymer is not fractal when the temperature of solution is low enough. At low temperature, the polymer is densely condensed into a ball which is called a globule. A phase transition occurs at a critical temperature, the configuration of the polymer becoming fractal above that temperature. This phase transition is known as the coil–globule transition.

2.4.6 *Phase transition of percolation*

In the vicinity of the critical point of a phase transition, many macroscopic quantities, such as specific heat and magnetisation, are known to follow power laws. Just at the critical point, these quantities or their derivatives are divergent. From a microscopic viewpoint, such behaviour is due to the divergence of correlation length. Since the correlation length is a length which characterises the system, the divergence indicates that the system becomes scale-invariant. For example, water at the critical point between gas phase and liquid phase (647 K, 218 atm.) is an opalescent fluid containing every possible size of drops. (Opalescence is the property of matter randomly reflecting light of any wavelength.)

The structure of matter at a critical point is a typical fractal. Let us consider an experiment on the percolation problem which exhibits a phase transition with a very simple mechanism. The problem of percolation occurs when fine metal particles are located at random on an insulator. If the covering of metal is very sparse it is not conductive. It becomes a conductor if the ratio is nearly equal to 1 – the ratio equals 1 when the surface of the insulator is completely covered by metal. There is a critical ratio, p_c, such that the layer behaves as an insulator if $p < p_c$ and as a conductor if $p > p_c$. This is a kind of phase transition from insulator phase to conductor phase and is called a *percolation transition*.

In Figure 2.11, micrographs of metal configurations are shown for five different values of p [23]. The largest clusters appear as dark shading in each picture. Figures 2.11(a) and (b) belong to the insulator phase because there is no percolating cluster, and (e) obviously corresponds to the conducting phase. In (c) and (d) we are near the critical point and can find large clusters. Rigorously speaking, the critical point, p_c, is defined as the smallest value of p at which there is a percolating cluster with probability 1 in an infinitely large system. The value of p_c estimated by this experiment is about 0.75.

As mentioned before, clusters at the critical point are characterised by a fractal dimension. This is confirmed by calculating the total mass, $M(r)$,

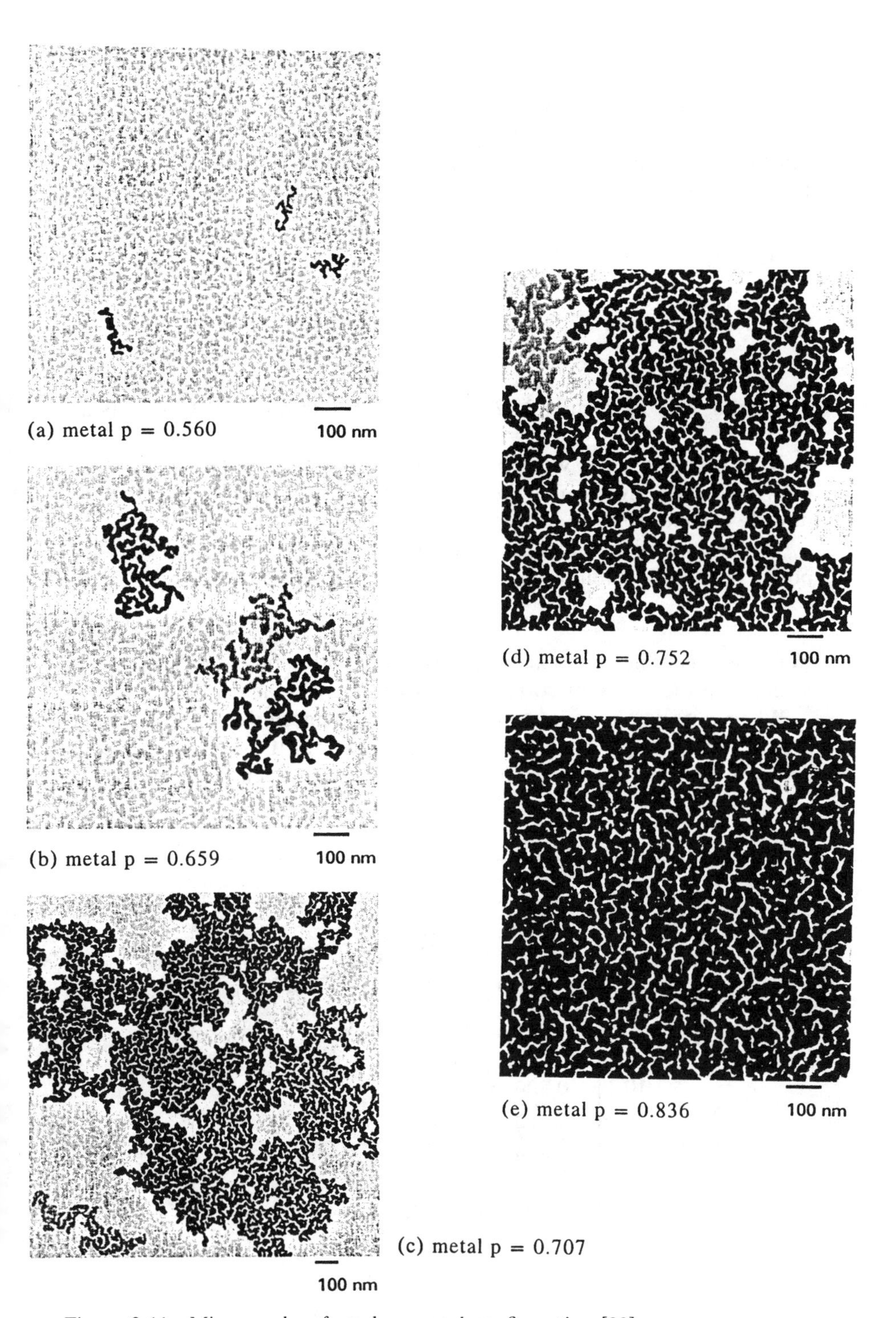

Figure 2.11 Micrographs of random metal configuration [23].

of a connecting metal cluster within a radius r (Equation (1.16)). Figure 2.12 shows a plot on a log-log scale of the single-cluster correlation function $G_c(r)$, that is, the probability that a point a distance r away is in the same cluster. Since $G_c(r)$ is directly related to the mass distribution by

$$M(R) \propto \int_0^R G_c(r) \cdot r \, dr, \tag{2.11}$$

the fractal dimension D is found directly from the exponent η_c of $G_c(r)$:

$$D = 2 - \eta_c. \tag{2.12}$$

The solid line in Figure 2.12 shows $\eta_c = 0.097$at $p = 0.752$, giving $D \doteqdot 1.9$.

In statistical physics, we define several exponents called critical exponents. They play an essential role in describing critical behaviour and are directly related to the fractal dimension.

We have mentioned that the correlation length ξ diverges at the critical point. It obeys the following power law near the critical point:

$$\xi \propto |p - p_c|^{-\nu}, \tag{2.13}$$

were ν is called the *correlation exponent*. The correlation length is defined to describe the decay of correlations over long distances.

When $p > p_c$, there is an infinitely large connected cluster. Another critical exponent, β, is defined via the cluster's size. If the fraction of area occupied by the infinite cluster be denoted by P_∞, then it satisfies

$$P_\infty \propto (p - p_c)^\beta. \tag{2.14}$$

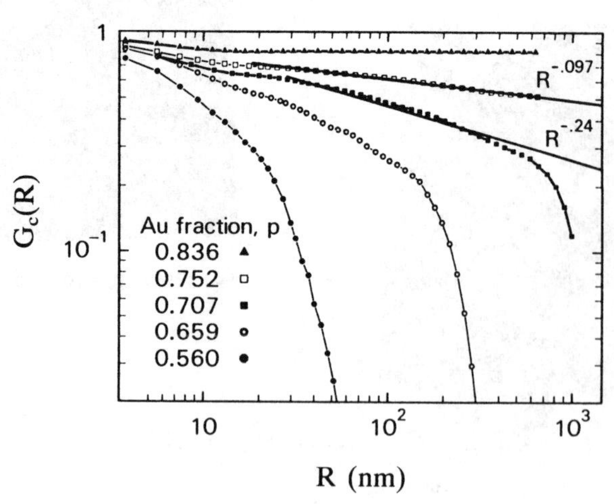

Figure 2.12 The single-cluster correlation function [23].

These critical exponents are related to the fractal dimension in the following way [24]. By using the above relations, the area, M, of infinitely large, connected cluster in an $L \times L$ square is given as

$$\begin{aligned} M &= p_\infty \cdot L^d \\ &\propto (p - p_c)^\beta \cdot L^d \\ &\propto \xi^{-\beta/\nu} \cdot L^d. \end{aligned} \tag{2.15}$$

Since M is a function of ξ and L, we denote it by $M(L, \xi)$. The proportionality coefficient, m, must be determined by the ratio of L to ξ because ξ is the only characteristic length:

$$M\,(L, \xi) = m\left(\frac{L}{\xi}\right) \cdot \xi^{-\beta/\nu} \cdot L^d. \tag{2.16}$$

In the limit $p \to p_c$, $M(L, \xi)$ must become a function of L only since ξ is divergent. In order to cancel the ξ dependence, $m(x)$ has to be proportional to $x^{-\beta/\nu}$, and we have

$$M(L, \infty) \propto L^{-\beta/\nu+d}. \tag{2.17}$$

This relation is directly related to the definition of the fractal dimension, for it says that the area of the cluster in a square of size L is proportional to $L^{d-\beta/\nu}$. Hence we finally obtain the fractal dimension D as

$$D = d - \frac{\beta}{\nu}. \tag{2.18}$$

The values of the critical exponents obtained by the above experiment are $\beta \doteqdot 0.14$ and $\nu = 1.35$, which are consistent with the previous result of $D = 1.9$.

For other phase transitions, such as magnetic systems, we can produce similar results. In each case the critical behaviour of macro-variables is governed by the fractal properties of underlying microscopic clusters.

2.4.7 *Turbulence*

A universal law of fluid dynamics is the *Reynolds similarity law* which says that the dynamical behaviour of two fluids with identical Reynolds numbers are similar, independent of their constituent molecules. Here the *Reynolds number* is defined by the equation

$$R = \frac{\rho U L}{\nu} \tag{2.19}$$

where ρ and ν denote mass density and viscosity, respectively, L specifies a characteristic length of the system, and U gives a characteristic fluid velocity. This law suggests that the Reynolds number is sufficient to define the dynamics of the fluid.

When the Reynolds number is small, viscosity stabilises the flow. On the other hand when it is greater than 10^4, the flow is unstable and becomes turbulent. For water at room temperature, ν is about $10^2\,\mathrm{s/cm^2}$, hence flow becomes turbulent for relatively small L and U – for example, $L \doteqdot 10\,\mathrm{cm}$ and $U \doteqdot 10\,\mathrm{cm/s}$. Nearly the same estimate can be made for air. Indeed, most of the water and air around us is in turbulent states. The complexity of the shape of cigarette smoke is due to turbulence.

Since the Reynolds number is proportional to the characteristic length of the system, it is expected that the characteristic length would vanish as

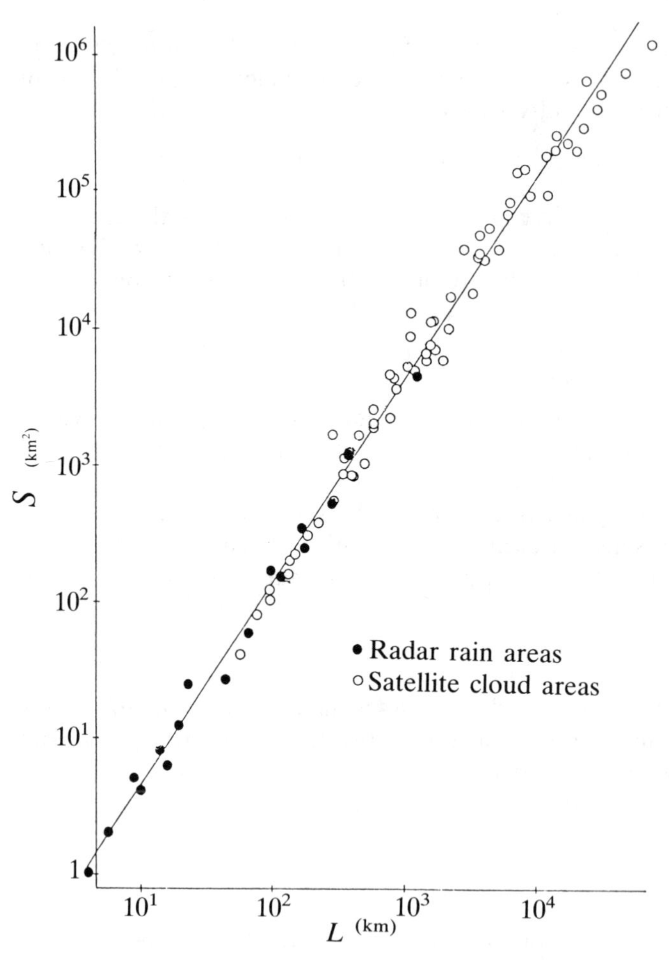

Figure 2.13 Area (S) and perimeter length (L) of clouds [26].

$R \to \infty$, and that an ideal turbulence with infinite Reynolds number would have fractal properties.

Experiment has confirmed that the energy dissipation region of isotropic turbulence in 3-dimensional space has a fractal structure. The correlation of energy dissipation rate, $\varepsilon(x)$, at x is given by [25]:

$$\langle \varepsilon(x) \cdot \varepsilon(x + r) \rangle \propto r^{-\mu}, \quad 0.2 < \mu < 0.5. \tag{2.20}$$

Here, energy dissipation means that kinetic energy of fluid is irreversibly transformed into thermal energy, and the rate is proportional to the square of the curl of the velocity field. Hence, (2.20) indicates that the velocity field has fractal properties. The fractal dimension of the energy dissipation region is defined by the exponent μ as

$$D = 3 - \mu. \tag{2.21}$$

Hence the dimension of turbulence is about 2.6. No theoretical approach has succeeded in deriving the fractal dimension directly from the Navier–Stokes equation. Turbulence is still a largely unresolved problem.

The shapes of clouds are fractals. In Figure 2.13, areas and perimeter lengths of various clouds as observed from a satellite are plotted on a log-log scale. The points are nearly on a straight line, and the slope gives the fractal dimension for the cloud shapes as 1.35 (Equation (1.14)) [26].

Figure 2.14 shows an example of *sumi-nagashi*, a turbulent flow pattern executed in indian ink, similar to marbling. It is produced in the following way: 'Agitate the water in a bowl and put a drop of indian ink into it. Then cover the surface of the water with a sheet of paper and pull it up after a few seconds.' We can easily confirm that the fractal dimension of *sumi-nagashi* is about 1.3. The scale of a cloud is nearly 10^6 times greater than that of *sumi-nagashi*, but the similarity of the shapes shows the universality of turbulence.

The twinkling of stars or distant lights is caused by fluctuations of the refractive index of the atmosphere. The refractive index is determined by density and humidity, and these quantities are shaken up by turbulence. Thus the twinkling is caused by turbulence.

There have been attempts to measure the fractal dimension of turbulence by observing the twinkling of light. It is argued [27] that the variance of fluctuation of light from distance L, $\langle \chi^2 \rangle$, should follow the relation

$$\langle \chi^2 \rangle \propto L^{(14-D)/6}, \tag{2.22}$$

where D is the fractal dimension of turbulence. By observing the change of $\langle \chi^2 \rangle$ over a range of L from 250 m to 2000 m, the dimension has been roughly estimated to be 2.5. This value is consistent with the preceding estimation.

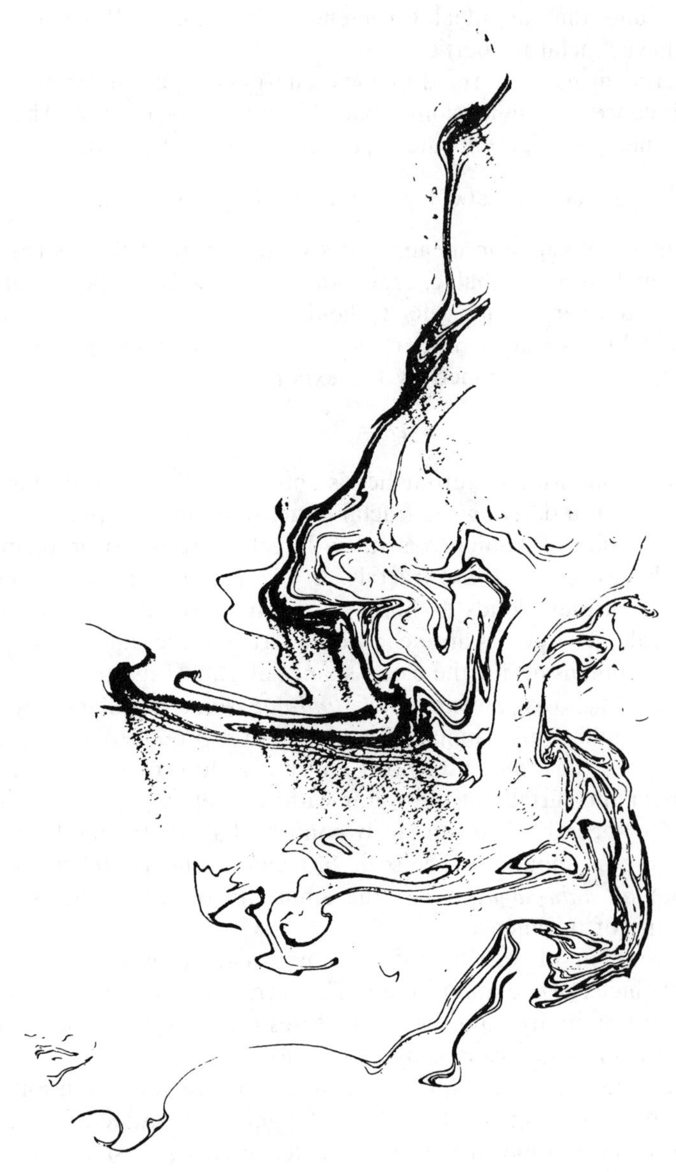

Figure 2.14 *Sumi-nagashi* (flow pattern produced by ink).

2.4.8 *Random walks*

In 1913 Perrin pointed out in his book *Atoms* [28] that the trajectory of a *Brownian motion* – the motion of microscopic particles suspended in a fluid – looks non-differentiable. It has been shown mathematically [1] that the fractal dimension of ideal Brownian motion is 2, hence the trajectory is everywhere non-differentiable and covers a plane completely. However, we may wonder whether high enough resolution would show the tajectory as connected line segments. In fact this is true. Physically the fractal dimension is valid for length scales larger than the mean free path. (See Section 6.1 for a detailed discussion.)

Brownian motion is sometimes compared with quantum-mechanical motion. Actually in some cases the Schrödinger equation can be deduced from Brownian motion [29], and as expected from this, the fractal dimension of the path of a particle in quantum mechanics has been shown to be 2 [30].

2.4.9 *Relaxation (amorphous and polymer)*

Relaxation of complicated materials often shows temporal fractal properties. A silk thread, which has been stretched under a load, starts contracting if the load is removed. In the case of simple relaxation a thread would contract exponentially to the original length. However, a silk thread does not contract so quickly. Instead it relaxes slowly following a power law, $t^{-\gamma}$. This experiment was first carried out by Weber and Gauss about 150 years ago, and has been repeated many times since then. This phenomenon of slow relaxation following a power law is called a *long time tail*.

Long time tails not only occur in mechanical relaxation but also in other phenomena, such as electric relaxation. The charge contained in a Leyden jar (a condenser using glass as the dielectric) also decreases in the form $t^{-\gamma}$. This relation has been confirmed to be valid to at least 16 million seconds (more than 180 days), and may be reasonably considered to hold indefinitely.

The exponent γ is not universal but depends on the substance. For some materials, this exponent is considered to characterise their amorphous structure.

Although long time tails have a long history, theoretical explanation has made little progress as yet. Since most materials with amorphous structures show the long time tails, it is expected that random fluctuation in spatial structure is playing an important role in producing the long time tail. (See Section 4.2 for theoretical approaches.)

Another type of relaxation which is closely related to fractals is the relaxation of a polymer in an electric field [33]. The experimentally obtained relaxation function $\phi(t)$ has the following form:

$$\phi(t) = \exp\left[-(t/T)^{\alpha}\right], \quad 0 < \alpha \leq 1. \tag{2.23}$$

The normal relaxation function is just the special case where $\alpha = 1$. However, actual values range between 0.3 and 0.8. The relaxation function is related to the complex dielectric constants as follows:

$$\varepsilon' + i\varepsilon'' = -\int_0^\infty e^{i\omega t} \frac{d\phi(t)}{dt}\, dt. \tag{2.24}$$

Combining these equations we have the power law

$$\varepsilon''(\omega) \propto \omega^{-\alpha}. \tag{2.25}$$

More rigorously, it is represented by the stable distribution defined in Section 5.2 as

$$\varepsilon''(\omega) = \pi T \omega p(\omega; \alpha, 0). \tag{2.26}$$

If we decompose the relaxation by a superposition of elementary relaxation functions in the form

$$\phi(t) = \int_0^\infty \lambda(\mu, \alpha)\, e^{-\mu t/T}\, d\mu, \tag{2.27}$$

the weight function $\lambda(\mu, \alpha)$ is given by the one-sided stable distribution,

$$\lambda(\mu, \alpha) = p(\mu; \alpha, -\alpha). \tag{2.28}$$

As will be discussed in Section 5.2, fractal and stable distributions are closely related and so the relaxation of polymers is also closely related to fractals.

2.4.10 *The Josephson junction*

In recent years high-temperature superconductors have been attracting the attention of both physicists and engineers. A Josephson junction is a junction that weakly connects two superconductors. In a superconductor, the phase of the electron wave function is coherent. When two superconductors are very close to each other, the wave functions in the superconductors show interference and the current i depends on the phase difference $\delta\phi$. Denoting the voltage between the superconductors by V, these variables satisfy the following equations:

$$i = i_c \sin \delta\phi, \tag{2.29}$$

$$V = \frac{\hbar}{2e} \frac{d}{dt} \delta\phi, \tag{2.30}$$

where i_c is the critical current, $\hbar$ is Planck's constant over 2π, and e is the electron charge. When a constant voltage is applied between the superconductors the phase difference $\delta\phi$ is integrated as

$$\delta\phi = \frac{2e}{\hbar} Vt + \text{const.}, \tag{2.31}$$

and the current becomes alternating with frequency

$$f = \frac{2|e|V}{\hbar}. \tag{2.32}$$

This frequency is usually in the microwave range. The curious behaviour of an alternating current flowing for direct voltage, is due tö quantum-mechanical effects.

A much more interesting phenomenon appears when an external microwave is applied to the junction. The microwave induces a direct current I in the circuit. The plot of current against voltage is given in Figure 2.15. Current does not increase linearly, but jumps at the following voltage:

$$V = \frac{\hbar}{2e} \cdot f \cdot \frac{P}{Q}, \tag{2.33}$$

where P and Q are prime integers. For small Q, the jump is larger, and the diagram becomes a devil's staircase [34].

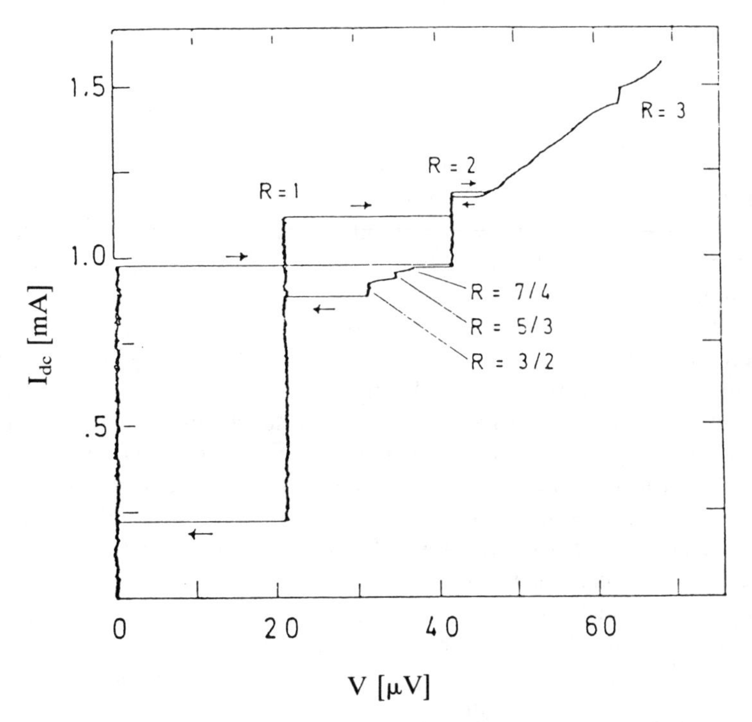

Figure 2.15 Current (I) v. voltage (V) in a Josephson junction [34].

2.4.11 *Molecular spectra*

Laser spectroscopy has revealed the very fine structure molecular spectra. It is found that in some cases the spectrum has a fractal structure as follows. A thick line spectrum under low resolution becomes several thin lines under high resolution, and each of the thin lines turns out to be composed of several thinner lines in even higher resolution. This indicates that the energy levels of a molecule have a fractal structure, since each line spectrum corresponds to an energy level.

A fractal distribution of energy level is expected to be connected with the problem of chaos. As will be seen in Section 3.2, a chaotic trajectory of a classical-mechanical system makes a fractal structure in state space. It might be expected that the energy level distribution of the corresponding quantum-mechanical system be a fractal. This problem of quantum chaos is a very basic problem in quantum mechanics. For example, even in the simplest system (that of one hydrogen atom in a uniform magnetic field [35]) the energy level has a fractal-like fine structure. It is known that the same system in classical mechanics shows chaotic behaviour. However, not all quantum-chaotic systems show fractal properties; a classical system with a fractal orbit does not always produce a quantum system with a fractal distribution of energy levels.

2.5 **Other topics**

2.5.1 *Flicker noise*

Fluctuations with power spectra of the form $f^{-\alpha}$ with α close to 1, are variously referred to as 'flicker noise', '$1/f$ noise', or 'pink noise'. This type of fluctuation was discovered about 60 years ago and examples have been found in many fields since then. However, no satisfactory theoretical explanation has yet been developed.

We learn in elementary statistical physics that the power spectrum of thermal noise is proportional to f^0. However, this has been found by experiment to be incorrect for small values of f. In most electrical circuits, a $1/f$ noise is observed at small f. This power spectrum indicates that the noise has a strong correlation with time. Intuitively this may be hard to accept, but it has been confirmed that the form of the spectrum does not change regardless of how long the observation period might be [36]. Hence we should regard it to be valid right to the limit of $f = 0$. This is analogous to the case of the long time tail.

The $1/f$ noise also occurs in the oscillation of crystals, heart beats, and the voltage fluctuation of nerve cells [37]. Most music has a $1/f$ noise component in its power spectrum. Such universality may indicate that the $1/f$ noise is produced by a simple mechanism, as yet unknown. It is likely to be closely related to fractals since the $1/f$ noise has self-similarity.

2.5.2 *Errors in communication*

The distribution of errors in data transmission lines is found to follow a power law [1]:

$$P(\tau) \propto \tau^{-D}, \quad D \doteqdot 0.3, \tag{2.34}$$

where $P(\tau)$ denotes the probability that the interval between two errors is larger than τ. This shows that the distribution of errors is a temporal fractal with dimension 0.3. Errors tend to form self-similar clusters like the Cantor set.

2.5.3 *Distribution of income*

It has been observed that annual income distribution is of the log-normal type with a power tail. The log-normal distribution is a distribution of positive numbers whose logarithms follow the normal distribution. In Figure 2.16 the distribution of income in the USA is plotted on log-normal graph paper for 1935–6 [38]. On such graph paper a cumulative log-normal distribution would be a straight line. That is the case for the first 98–9 percentiles; however, afterwards it follows a power law [39]:

$$P(X) \propto X^{-1.6}. \tag{2.35}$$

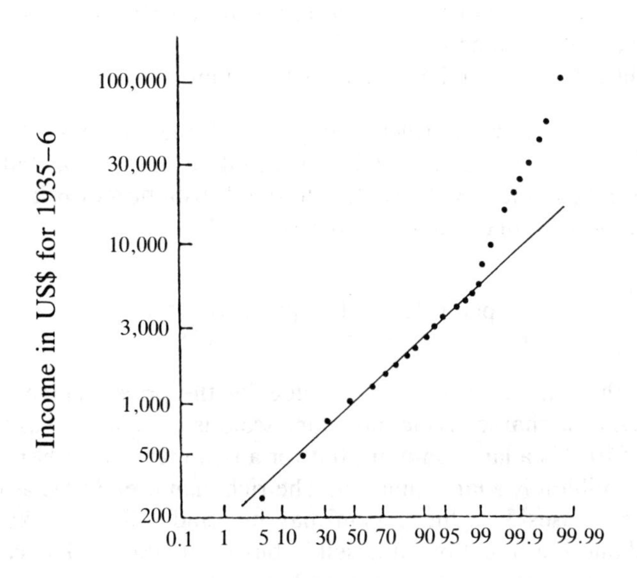

Figure 2.16 Distribution of income (in 1935) on log-normal graph paper. Abscissa shows cumulative percentage, e.g. income of about 90 people are less than $3000 [38].

That is, the income of rich people is distributed fractally.

The log-normal distribution may be described by the following process. Consider the probability of success of a primary task, P_0. If the task requires the successful completion of many independent subtasks, the success for the primary task becomes

$$p_0 = p_1 \cdot p_2 \cdot p_3 \ldots, \tag{2.36}$$

where p_i denotes the probability of success of the ith subtask. So taking logarithms of both sides we have

$$\log p_0 = \log p_1 + \log p_2 + \log p_3 + \ldots. \tag{2.37}$$

Since the p_i are independent random variables, the central limit theorem is applicable and $\log (p_0)$ has a Gaussian distribution.

The reason for the power tail is not clear.

2.5.4 *Stock price changes*

The change of stock prices seems to be quite random. Mandelbrot, however, found two laws in the randomness [1]:

(1) Price change in unit time is describe by a stable distribution with characteristic exponent $\alpha \doteqdot 1.7$.
(2) The distribution is independent of time unit.

The first law shows a fractal property of the process. As will be described in Section 5.2, the stable distribution has a long tail characterised by the exponent α. If we denote the distribution of price change x in a unit time t by $p(x)$, then it satisfies

$$\int_x^{\infty} \mathrm{p}(x')\,\mathrm{d}x' \propto \int_{-\infty}^{-x} \mathrm{p}(x')\,\mathrm{d}x' \propto x^{-\alpha}. \tag{2.38}$$

Therefore there is no characteristic value for the price change.

The lack of a characteristic monetary scale is recognised easily. For a poor man \$1000 is a large amount, but for a rich man it is rather little and perhaps \$1 million is a large amount. The rich man uses \$1000 as easily as the poor man uses \$1. One person may buy and sell in blocks of 1000 stocks and another might buy and sell in blocks of 100 000. Hence there is no characteristic amount in money and stock price.

The second law indicates that the change of stock price is fractal in time, that is, the graph of stock price change in one day becomes statistically identical to that of the change in a year if we rescale the axes appropriately.

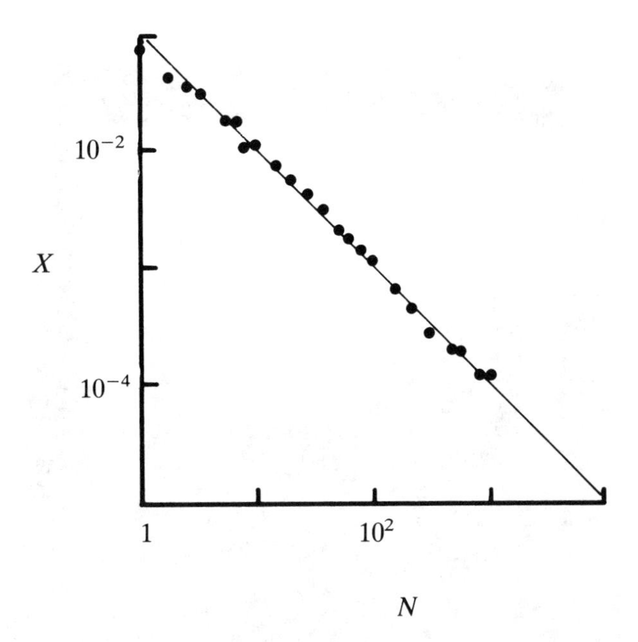

Figure 2.17 Frequency of words (X) and the order (N) in log-log plot.

2.5.5 *Zipf's law*

If we count the frequency of words in a piece of text we obtain a plot as shown in Figure 2.17. Here X denotes the probability of appearance and N specifies the words in order of frequency. For example, $N = 1$ might correspond to 'the' and $N = 2$ to 'of'. The points in the figure are almost on a straight line of gradient 45°. As mentioned in Section 1.3, this distribution is equivalent to

$$P(X) \propto X^{-1} \tag{2.39}$$

where $P(X)$ denotes the probability that a randomly chosen word's probability of appearance is larger than X. This is a fractal distribution with the exponent $D = 1$.

This type of distribution is called *Zipf's law*. It has been found in many other fields such as, the number of cities ranked by population. As with $1/f$ noise there is no satisfactory theoretical explanation in spite of the long history.

2.5.6 *Video feedback*

When a video camera is focused on its monitor screen, unpredictably complicated phenomena occur [40]. A picture obtained in this way is

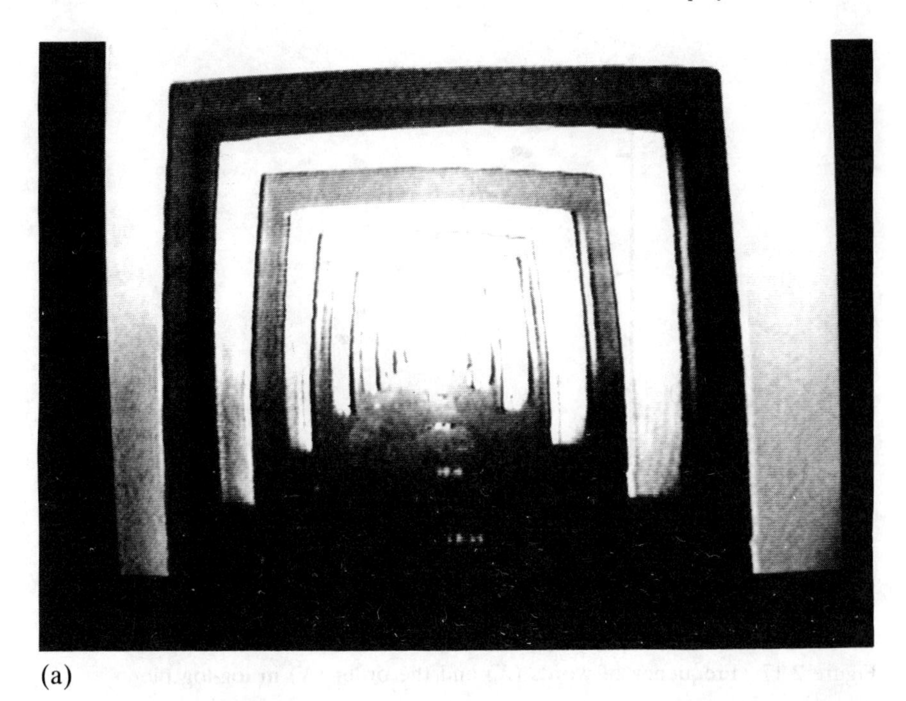

(a)

(b)

Figure 2.18 (a) Video feedback; (b) strange shapes appear by zooming up.

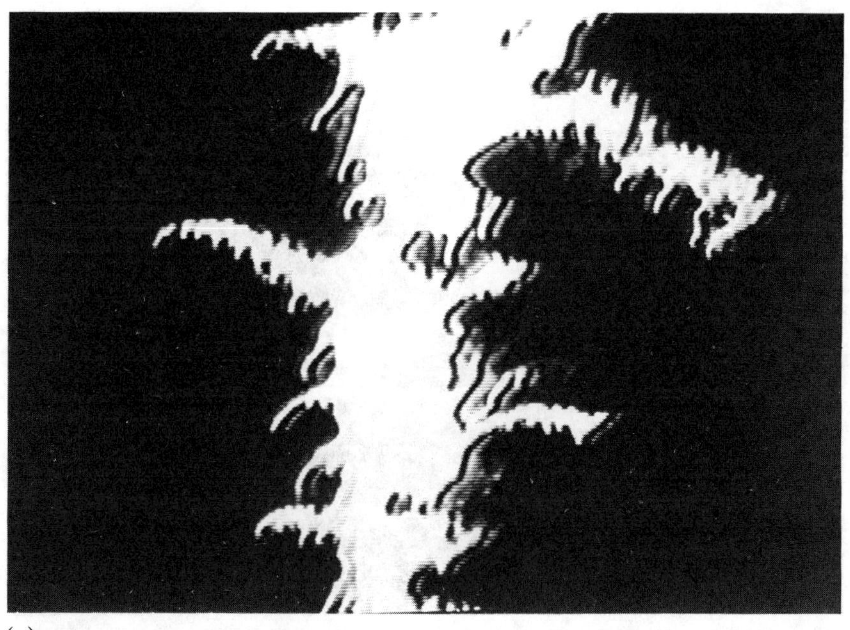

(a)

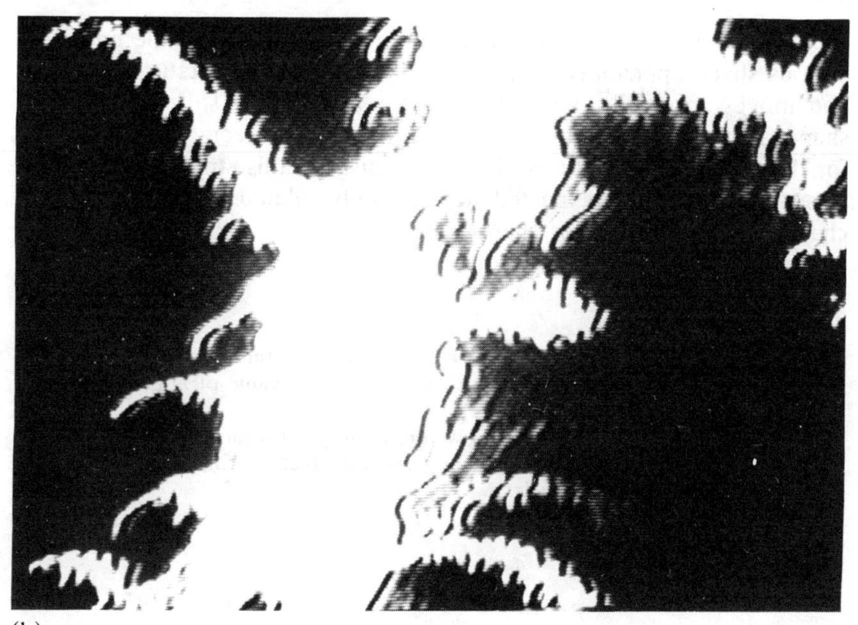

(b)

Figure 2.19 Fractals in video feedback (three successive snapshots).

(c)

shown in Figure 2.18(a). But if we zoom up the centre of the images, a curious shape appears (Fig. 2.18(b)). The shape is not static but rotates and moves about like an animal. If we zoom it up still more, fractal shapes appear to gush out from the centre of the screen; see Figure 2.19 for three successive pictures. The mechanism of this strange phenomena is not simple, and is expected to be closely related to the problem of chaos (see Section 3.2 for chaos).

Notes

1 In Figure 2.4 the points are shown on a straight line, but note that these points are subject to experimental error. The error in the value of D is estimated to be about 10%.

2 Some quantum chaos systems show fractal properties but others do not. The problem is not so simple that the classical fractal orbit does not always produce fractal energy levels.

3 Fractals on the computer

In this chapter we consider computer-simulated fractals. All the fractals discussed here are produced by simple rules. We shall imitate the simplest fractals in nature to find the essence of formation of fractal structures; we can also make many artificial fractals which may not exist in nature. Readers who have access to a computer are recommended to create fractals on them – program listing are given in Section 3.7.

3.1 Aggregation

We have seen in Section 2.4 that aggregations of fine particles make fractal structures. Such structures can be produced easily by simulating the aggregation process on a computer.

First, place a 'particle' at the origin as a 'seed'. Then let another 'particle' move randomly starting from a point far from the 'seed'. When the moving particle comes close and touches the seed, it stops moving and is considered to be forming part of an aggregation. Introduce another particle in the same way and allow it to follow a random walk until it touches the growing cluster and becomes a part of the aggregate. Repeat this process many times over (Fig. 3.1).

An example of the resulting cluster is shown is Figure 3.2. Because a randomly walking particle is likely to attach to peripheral parts, the structure becomes dendritic. (We use the word 'dendritic' in its original meaning in mineralogy and neurophysiology – not in the narrower meaning used more recently in solid state physics.) Although this aggregate is constructed on a 2-dimensional lattice, it looks very similar to the real aggregates shown in Section 2.4. Since this process of aggregation is governed by diffusion, it is called *diffusion-limited aggregation* (DLA) [41].

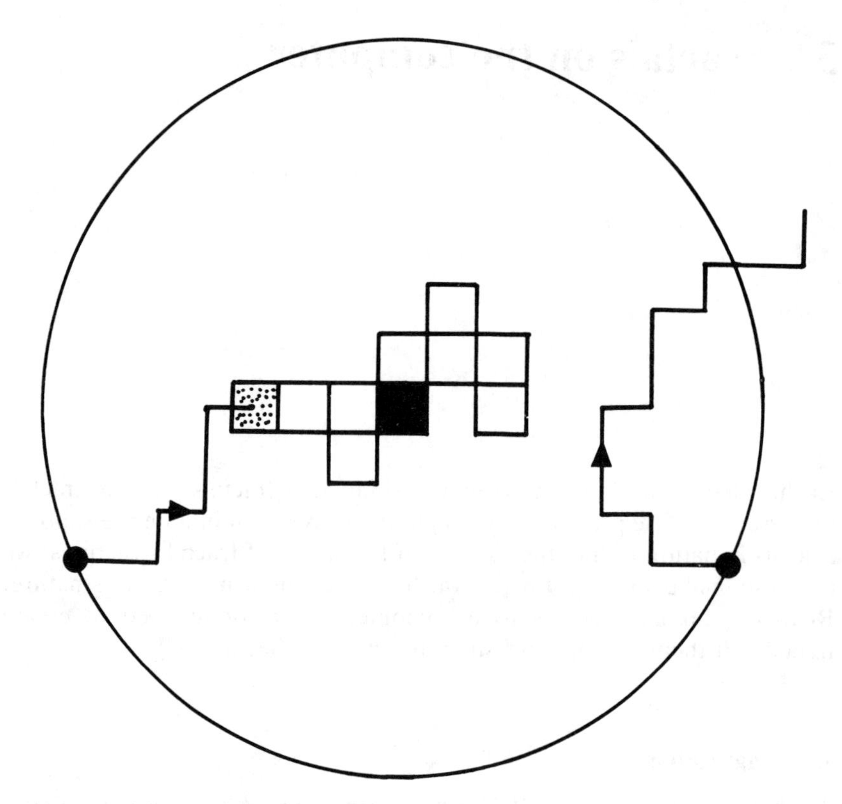

Figure 3.1 Simulation of aggregation.

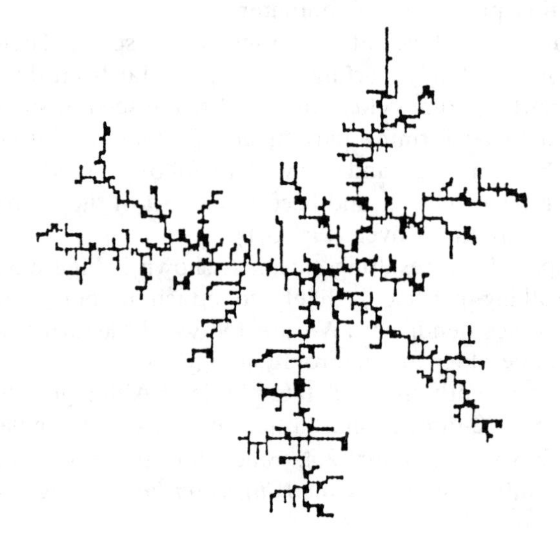

Figure 3.2 An example of aggregation.

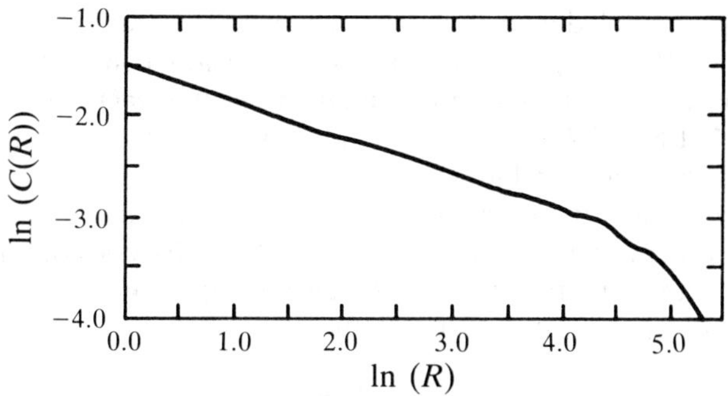

Figure 3.3 Correlation function in log-log plot [42].

In Figure 3.3, the correlation function is plotted as a function of distance on a log-log scale. Clearly the correlation decays according to a power law. The exponent of the power determines the fractal dimension of the aggregation as D $\doteqdot$ 1.66, for $d = 2$ [42]. This value agrees very well with that of aggregation in nature, 1.7.

For DLA, the following relations have been proposed between space dimension d and the fractal dimension D [42],

$$D = \frac{5}{6}\, d, \tag{3.1}$$

or [43],

$$D = \frac{d^2 + 1}{d + 1} \tag{3.1'}$$

Theoretical arguments in favour of (3.1′) are given in [44] and [45]. For $d = 2$ and $d = 3$, the values of D predicted by these formulae are identical and very close to experimental values.

In an early stage of research, it was believed that the structure and the fractal dimension of DLA were universal, that is, they were believed to be independent of lattice structure. However, when clusters with as many as 4×10^6 particles were simulated, the anisotropy of the underlying lattice was apparent [46]. In that case, the structure of aggregation cannot be specified by only one parameter, the fractal dimension, and more complicated treatment is needed (see [47] for details).

In the above model, aggregation grows from a fixed cluster seed. A generalised version of this model is the so-called *cluster–cluster aggregation* model (CCA), in which a number of particles, initially located at random, start moving randomly. When two particles collide they adhere and form a cluster. Clusters move and adhere in the same way and they grow bigger and bigger.

The structure of aggregation obtained by this method is, as expected, fractal. The fractal dimensions are, however, not identical to those of DLA. Also, they depend on how the diffusion constant depends on cluster size. In 2-, 3- and 4-dimensional space, the fractal dimensions are roughly 1.4, 1.8 and 2.0, respectively [48]. These values are much smaller than the corresponding DLA values – 1.66, 2.5 and 3.3. In the CCA model, larger gaps between branches are created when clusters collide, hence the structures become sparser than DLA. By considering a simple geometrical model, the fractal dimension of a CCA cluster in d-dimensional space is estimated as [49]:

$$D = \frac{\log(2d + 1)}{\log 3}. \tag{3.2}$$

3.2 **Chaos and maps**

3.2.1 *Strange attractors*

A very interesting phenomenon occurs in the solution of the following set of nonlinear differential equations called the *Lorenz system*:

$$\begin{aligned} \frac{\mathrm{d}}{\mathrm{d}t} x &= -10\,(x - y), \\ \frac{\mathrm{d}}{\mathrm{d}t} y &= -xz + r\,x - y, \\ \frac{\mathrm{d}}{\mathrm{d}t} z &= xy - \frac{8}{3}\,z. \end{aligned} \tag{3.3}$$

This system arises from problems related to fluid convection and to weather forecasting. When the parameter r lies in the interval $24.7 < r < 145$, the solution does not converge to a fixed point in the limit $t \to \infty$ nor is there a limit cycle, but the solution keeps moving around in a finite region [50]. Figure 3.4 shows an example of solution in phase space. The orbit might look overlapping in the figure, but it never overlaps in xyz-space because the solution has been proved to be unique.

The power spectrum of this motion is continuous which indicates that the motion of the point (x, y, z) is not periodic but is very complicated. Furthermore it is known that this system is unstable in the sense that the distance between two neighbouring points in xyz-space increases exponentially with time. Even if the initial difference between two initial states is negligibly small, this difference grows very rapidly and ceases to be negligible after a while. Such complicated non-periodic motions in deterministic systems are called *chaos*. The change of weather is expected to be chaotic, hence weather forecasting is very difficult.

In contrast to the distance between two points, the volume of a small cube in xyz-space decreases exponentially with time, since the volume V,

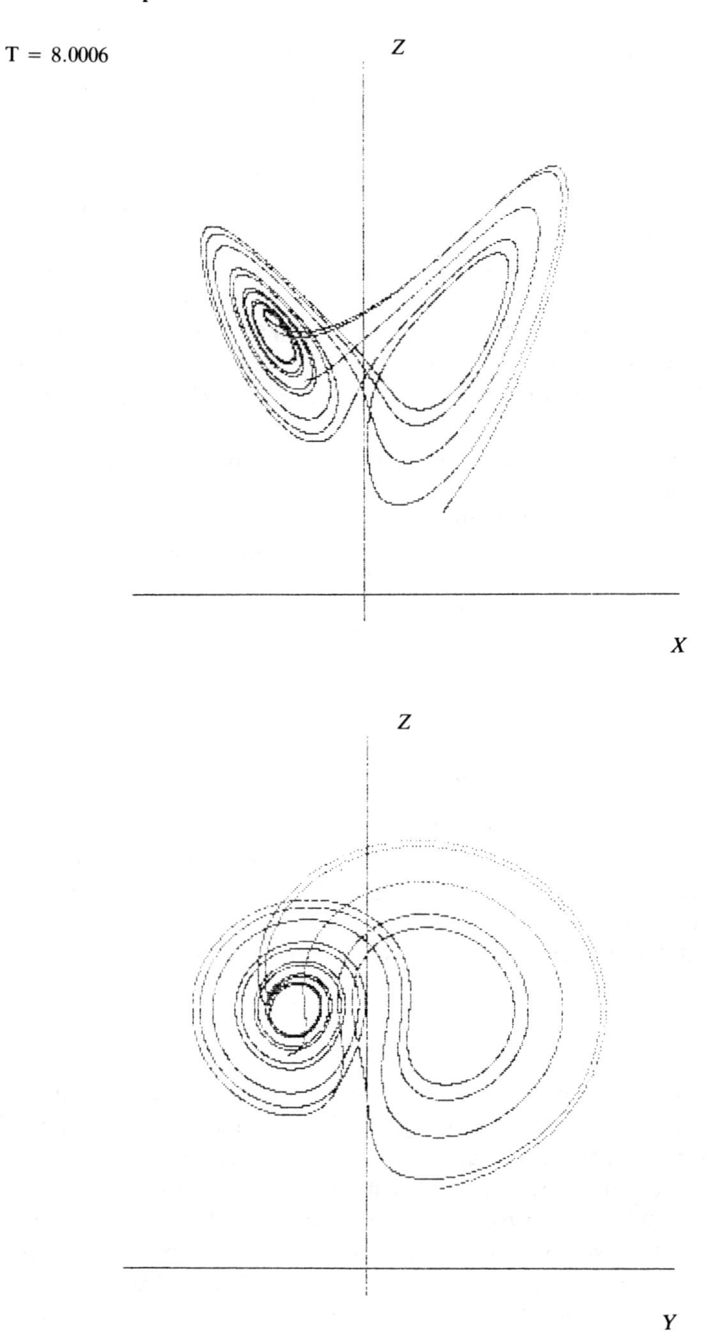

Figure 3.4 The trajectory of the Lorenz system.

if it is sufficiently small, satisfies the following equation:

$$\frac{1}{V}\frac{dV}{dt} = \frac{\partial}{\partial x}\frac{dx}{dt} + \frac{\partial}{\partial y}\frac{dy}{dt} + \frac{\partial}{\partial z}\frac{\partial z}{\partial t} = -\frac{41}{3}. \tag{3.4}$$

Therefore, a cube must be deformed into a long thin ribbon in order to satisfy the above two conditions: distance between any two points increases while the volume decreases.

The limit set of the orbit at $t \to \infty$ is generally called the *attractor*. It has been confirmed numerically that the attractor of the Lorenz system has a very complicated structure with infinitely many foldings. The initial cube is thinned, elongated and folded over and over again indefinitely as $t \to \infty$. Such complicated attractors found in chaos are called *strange attractors*.

Strange attractors have been found in many systems with few degrees of freedom. In linear systems attractors are, of course, not strange. The following system (called the Rössler system) is famous for showing that chaos can be produced with only one nonlinear term (xz) [51]:

$$\begin{aligned}\frac{d}{dt}x &= -(y + z),\\ \frac{d}{dt}y &= x + 0.2y,\\ \frac{d}{dt}z &= 0.2 - 5.7z + xz.\end{aligned} \tag{3.5}$$

Attractors of ordinary differential equations with the degree of freedom less than 2 are limited to either a fixed point or a limit cycle, and have proved not to be strange. However, even in systems with only two variables, chaos can be found if the system evolves discretely. In Figure 3.5, the strange attractor is shown for the so-called *Henon map*:

$$\begin{aligned}x_{n+1} &= 1 - ax_n^2 + by_n,\\ y_{n+1} &= x_n.\end{aligned} \tag{3.6}$$

If we initially consider a small square, its area decreases exponentially, while the distance between a pair of points increases exponentially. As a result, the square is transformed into a thread-like structure (see Fig. 3.6). Since the attractor of this map is confined to a finite region, the thread is folded and stretched again and again. In Figure 3.5 diagram (b) magnifies the squared part of the attractor shown in (a); and (c) and (d) show further magnifications of the attractor. We find that a single line in low resolution becomes a bundle of lines in higher resolutions in a self-similar way. The complicated structure created by the folding and elongation is, as expected, a fractal.

Strange attractors in systems of ordinary differential equations also

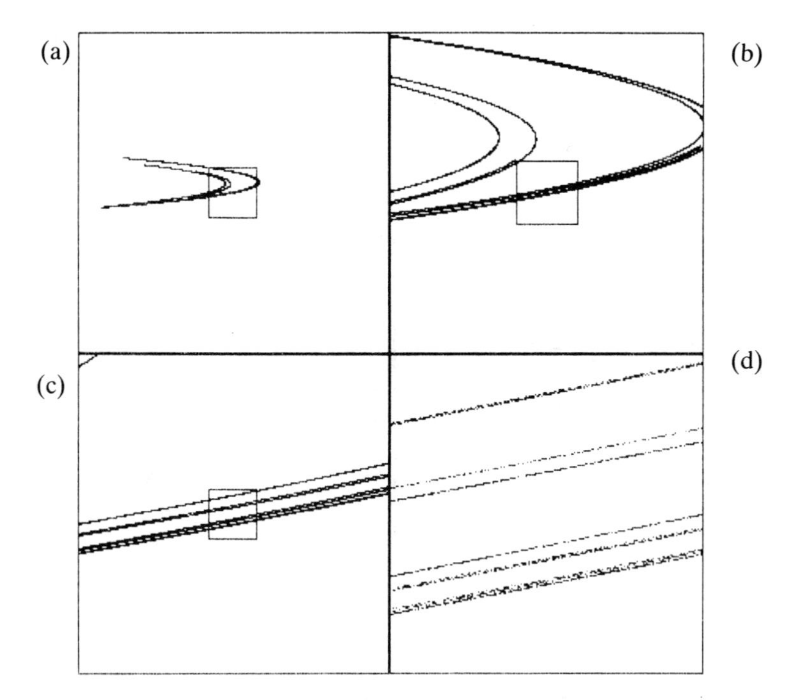

Figure 3.5 The strange attractor of the Hénon map: (a) the global structure; (b) magnification of the squared area in (a); (c) magnification of (b); (d) magnification of (c).

usually have fractal properties. By imagining a plane in the phase space and observing only the points where the orbits pass through the plane, the dynamical systems can be reduced to a discrete map called the *Poincaré map*. The Poincaré map of the Rössler system, like the Henon map is self-similar and we know that the Rössler attractor is also fractal. The latter is also true for the Lorenz attractor.

Once we know that strange attractors are fractals, we can calculate their fractal dimensions. Measured for example by the method of changing resolution, the fractal dimensions of the attractors of the Hénon map and the Lorenz system are estimated to be 1.26 and 2.06, respectively [52]. We can quantitatively characterise the complexity or strangeness of strange attractors by their fractal dimensions. The name 'strange' was given before the introduction of fractals, and it now seems more appropriate to call them 'fractal attractors' because their true character has been unmasked.

Can we theoretically calculate the fractal dimensions of attractors? Unfortunately, no formula has yet been found which gives the fractal dimension of an attractor directly from a given dynamical system, though there is an indirect formula which gives the fractal dimension from

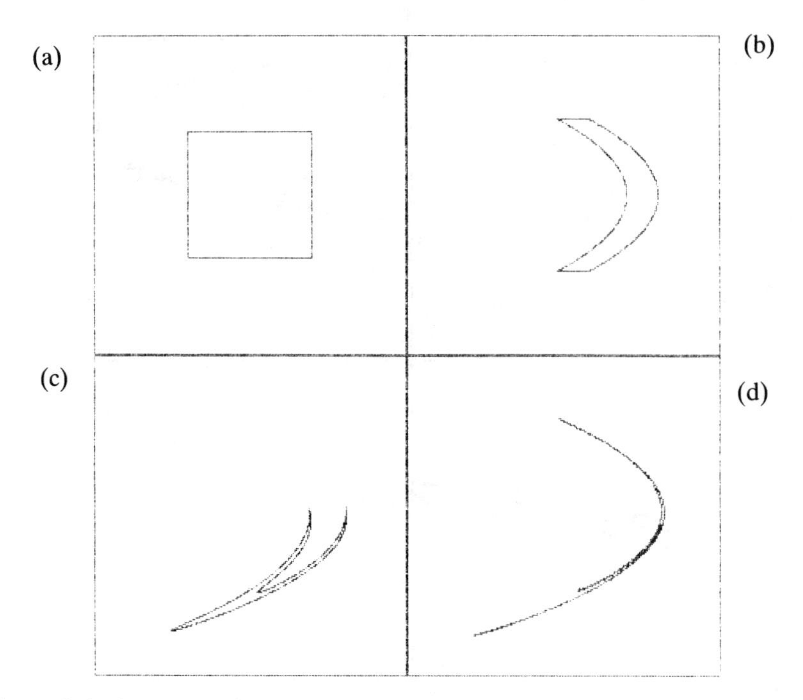

Figure 3.6 Sequence of transformations of initial square.

Lyapunov exponents. Lyapunov exponents λ_α are defined as follows. Consider two points aligned in the direction α and separated by a distance $L_\alpha(t)$ at time t. If we denote by $L_\alpha(t + \tau)$ the distance at time $t + \tau$, then the Lyapunov exponent is defined via the mean expansion rate as

$$\lambda_\alpha \equiv \frac{1}{\tau} \langle \log \frac{L_\alpha(t + \tau)}{L_\alpha(t)} \rangle. \tag{3.7}$$

If λ_α is positive (resp. negative) the two points separate (resp. approach) exponentially. The subscript α specifies the direction and we need as many αs as the number degrees of freedom. That is, in d-dimensional phase space, we need d Lyapunov exponents $\lambda_1, \lambda_2, \ldots, \lambda_j$. The fractal dimension of the strange attractor is then given by the following formula [53]:

$$D = j - \frac{\lambda_1 + \lambda_2 + \ldots + \lambda_j}{\lambda_j} \tag{3.8}$$

where we have put the subscripts in order of magnitude and j is the smallest integer which makes $\lambda_1 + \lambda_2 + \ldots + \lambda_j$ negative, namely

$$j \equiv \min \left\{ n \mid \sum_{i=1}^{n} \lambda_i < 0 \right\}. \tag{3.9}$$

In the case of the Hénon map and the Lorenz system, the Lyapunov exponents have been obtained numerically as $\lambda_1 = 0.42$, $\lambda_2 = -1.58$ and $\lambda_1 = 1.37$, $\lambda_2 = 0.00$, $\lambda_3 = -22.4$, respectively [54]. Substituting these values into (3.8), we obtain the fractal dimensions as 1.26 and 2.06, respectively. This agrees very well with the above-mentioned values calculated by other methods. The fractal dimension defined by (3.8) is called the Lyapunov dimension or Kaplan–Yorke dimension and in general it is equal to the information dimension of the strange attractor [55].[1]

3.2.2 *Chaos in maps*

In the last section we have seen that strange attractors are produced as a result of infinite repetition of foldings. Here we study the property of folding by considering a 1-dimensional map.

Let us consider a simple nonlinear map called the *logistic map*:

$$x_{n+1} = r \cdot x_n \cdot (1 - x_n), \quad 0 \leq r \leq 4. \tag{3.10}$$

As seen from Figure 3.7, the interval [0, 1] is folded once by one iteration, and this map may be regarded as the simplest model of folding.

The asymptotic behaviour of x_n depends strongly on r as follows:

(1) For $0 \leq r < 1$, x_n decreases as n and x_n approach 0.
(2) For $1 \leq r \leq 2$, x_n monotonically approaches $1 - (1/r)$.

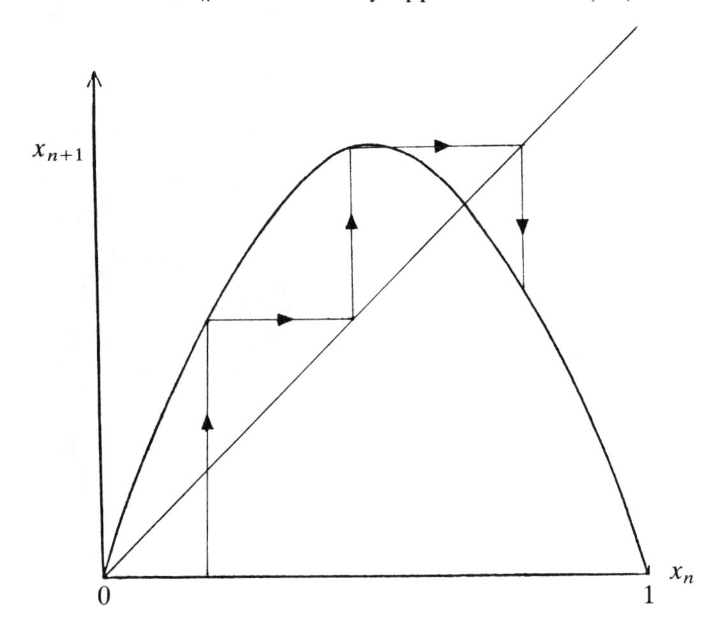

Figure 3.7 Logistic map.

(3) For $2 < r \leq 3$, x_n approaches $1 - (1/r)$ with oscillation.
(4) For $3 < r \leq 1 + \sqrt{6} = 3.449$, x_n is gradually approaches periodic motion of period 2.
(5) For $1 + \sqrt{6} < r \leq 4$, the behaviour of x_n is very complicated. The set of attractors of x_n is shown in Figure 3.8. For arbitrarily large n, and for fixed r, x_n keep moving on the black points. As r is increased periods of 4, 8, ..., 2^n appear successively. The period of 2^∞ appears at $r_c = 3.57$. At values of r where the dotted points in the figure line up densely, x_n moves 'chaotically'. The small white regions in the figure are called 'windows' and there x_n is periodic.

It is remarkable that such a simple 1-dimensional map can produce chaotic motion.

In the region where the periods 2^n appear successively Feigenbaum found two interesting relations [56]. First, if r_n is the lowest value at which a period of 2^n appears, then it was proved that as n tends to infinity,

$$\frac{r_n - r_{n-1}}{r_{n+1} - r_n} \to \delta = 4.669 \ldots \tag{3.11}$$

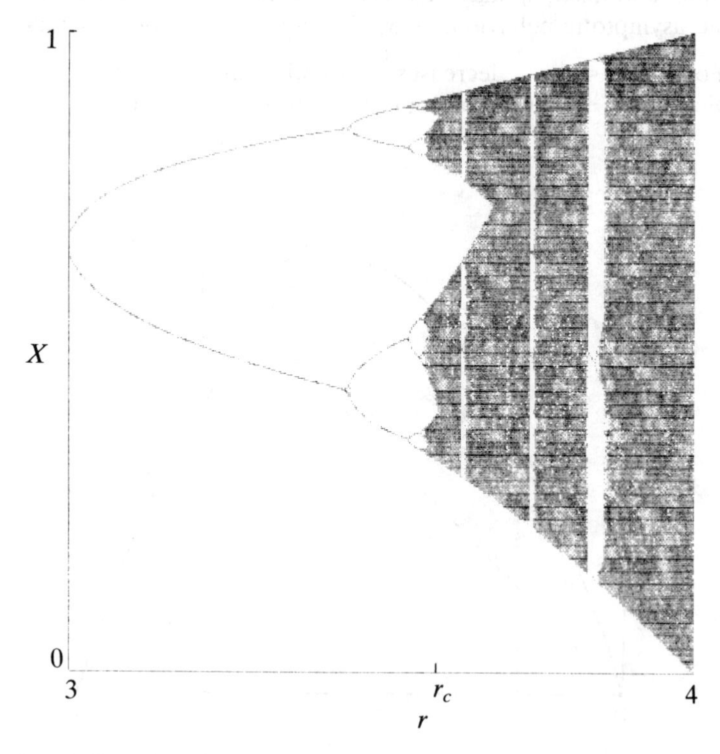

Figure 3.8 Bifurcation diagram of logistic map for $3 \leq r \leq 4$.

Secondly, for power spectrum of frequency $\omega=\omega_0/2^n$, the following relation holds:

$$\frac{S\left(\frac{\omega_0}{2^n}\right)}{S\left(\frac{\omega_0}{2^{n+1}}\right)} \rightarrow \mu = 6.57\ \ldots \tag{3.12}$$

The constants δ and μ introduced here prove to be universal, that is, the values defined by (3.11) and (3.12) are identical in any nonlinear system. This universality has also been confirmed experimentally [57].

A famous theorem about chaos in 1-dimensional maps is the *Li–Yorke theorem* [58]. It applies to any continuous map $f(x)$ from an interval I to itself such that there exist four points a, b, c and d such that

$$d \leq a < b < c \qquad \text{where } f(a) = b,\ f(b) = c \text{ and } f(c) = d.$$

Then the following statements can be proved:

(1) There exist k-periodic points ($f^k(x) = x$) for any natural number k.
(2) There exists an uncountable set S such that any two distinct points p and q satisfy the following relations:

$$\lim_{n\to\infty} \sup |f^n(p) - f^n(q)| > 0,$$
$$\lim_{n\to\infty} \inf |f^n(p) - f^n(q)| = 0. \tag{3.13}$$

(3) Let p be a point in S and q be a k-periodic point, then the following inequality holds

$$\lim_{n\to\infty} \sup |f^n(p) - f^n(q)| > 0. \tag{3.14}$$

The above condition for f is called a generalised condition for period 3 since the special case $d = a$ corresponds to the case of period 3. Statement (3) says that there exist uncountably many points which do not approach any periodic points, and (2) states a kind of ergodicity that those points come very close to each other but never converge. In the logistic map, period 3 appears at $r = 3.8284$, and the above theorem holds for r larger than that value.

Historically, the logistic map was obtained from the logistic equation, which describes the growth of a population in a closed area:

$$\frac{\mathrm{d}}{\mathrm{d}t} u = (\varepsilon - hu)u. \tag{3.15}$$

Making this equation into a difference equation,

$$\frac{u_{n+1} - u_n}{\Delta t} = (\varepsilon - hu_n)\, u_n, \tag{3.16}$$

we obtain the logistic map if we change the variables as

$$r = 1 + \varepsilon \Delta t,$$
$$x_n = \frac{h \Delta t}{1 + \varepsilon \Delta t} u_n. \tag{3.17}$$

The solution of (3.15) can be obtained analytically for any initial condition $u(0) > 0$. It monotonically approaches a fixed point ε/h. By contrast, the difference equation for large intervals Δt and the logistic map behave quite differently, producing chaos. This kind of discrepancy between the solution of a differential equation and that of its difference equation appears in any nonlinear system if the difference interval is sufficiently large. Hence we have to be careful when we numerically solve a differential equation by using a difference equation.

'*Symbolic dynamics*' is an interesting approach to the 1-dimensional map. In the case $r = 4$, the logistic map can be transformed through a transformation

$$y_n = \frac{2}{\pi} \sin^{-1} \sqrt{x_n} \tag{3.18}$$

into the following triangular map (Fig. 3.9):

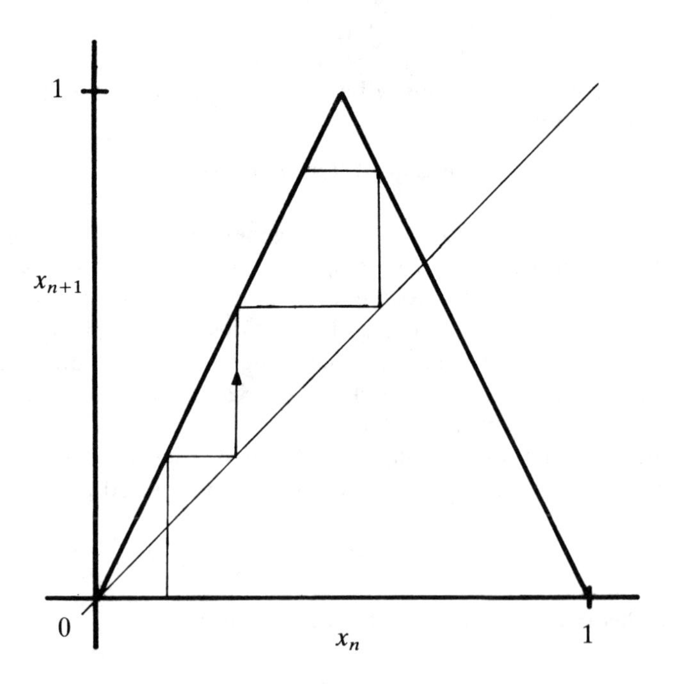

Figure 3.9 The triangular map.

$$y_{n+1} = \begin{cases} 2y_n & 0 \leq y_n \leq 0.5, \\ 2-2y_n & 0.5 \leq y_n \leq 1. \end{cases} \tag{3.19}$$

For any given initial value, y_0, the solution $y_1, y_2, \ldots, y_n, \ldots$ is determined uniquely. For this solution $\{y_n\}$, we construct a sequence of symbols $\{w_n\}$ consisting of 0s and 1s by the following simple rule: if $y_n > 0.5$ then $w_n = 1$, otherwise $w_n = 0$. For a given y_0, the sequence $\{w_n\}$ is determined uniquely, although the order of appearance of the symbols 1 and 0 looks quite random. Actually, it has been proved that the set of sequences $\{w_n\}$ is mathematically equivalent to the set of random sequences of 1 and 0 created by tossing a coin (head and tail correspond to 1 and 0, respectively). This randomness originates in the chaotic motion of x_n in the logistic map.

Let us now construct $\{w_n\}$ for a given y_0. It is convenient to associate to $\{w_n\}$ a real number in the interval [0, 1] in the following way:

$$z = \sum_{n=0}^{\infty} w_n \cdot 2^{-n-1}. \tag{3.20}$$

That is, we regard w_n as the $(n+1)$th digit in the binary expansion of z. Then the problem consists in determining z as a function of the initial value y_0. Here we introduce two transformations, f_0 and f_1:

$$f_0(z) = \frac{1}{2} z, \tag{3.21}$$

$$f_1(z) = \frac{1}{2} + \frac{1}{2} z. \tag{3.22}$$

By these transformations, the sequence $\{w_n\}$ is shifted as $w_n \rightarrow w_{n+1}$ for $n \geqslant 1$ while w_0 is 0 for f_0 or 1 for f_1. Thus the initial value y_0 is transformed as

$$f_0\colon y_0 \rightarrow \frac{1}{2} y_0, \tag{3.23}$$

$$f_1\colon y_0 \rightarrow 1 - \frac{1}{2} y_0. \tag{3.24}$$

Combining (3.21)–(3.24), we have

$$z\left(\frac{y_0}{2}\right) = \frac{1}{2} z(y_0). \tag{3.25}$$

$$z\left(1 - \frac{y_0}{2}\right) = \frac{1 + z(y_0)}{2}. \tag{3.26}$$

From these functional equations we can solve z for any given y_0. For example, by substituting $y_0 = 0$ into (3.25) and (3.26), we obtain $z(0) = 0$

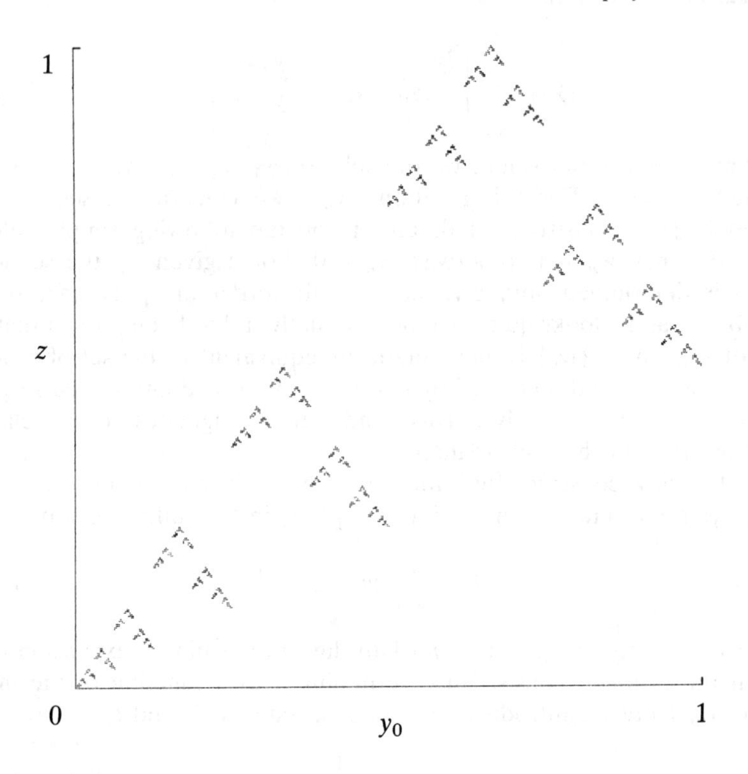

Figure 3.10 Initial value (y_0) and the symbolic solution (z) in the triangular map.

and $z(1) = \frac{1}{2}$. Substituting $y_0 = 1, \frac{1}{2}, \frac{1}{4}, \frac{3}{4}, \ldots$ we obtain the solution shown in Figure 3.10. The graph is an everywhere-disconnected fractal. Because it is not continuous, the values of z for two infinitesimally close initial values give a finite difference. Conversely, when a symbolic sequence $\{w_n\}$ is given, the initial value y_0 which gave the sequence can also be determined from this figure. Thus the triangular map has been solved completely.

3.2.3 *Fractals by maps*

In this section we give some examples showing that complicated fractal figures can be produced by simple maps.

For a given map

$$x_{n+1} = f(x_n) \tag{3.27}$$

the set of initial points $\{x_0\}$ whose iterated points never diverge ($|x_n| < \infty$ for any n) is called its *Julia set*. For many maps, the Julia sets are known to be fractals. A good example is the following complex logistic map:

$$f(z) = az(1 - z). \tag{3.28}$$

In Figure 3.11, the Julia set is shown for the case $a = 3.3$. It is surprising that such a complicated shape is hidden in the simple map (3.28).

Changing the origin and the scale of z, and defining $b = a^2/4 - a/2$, the above function can be written as

$$g(z) = z^2 - b. \tag{3.29}$$

The set of complex parameters b such that successive iterates of $z = 0$ under $g(z)$ do not tend to ∞ is named the *Mandelbrot set*. As shown in Figure 3.12 this set has a fractal border [1].

When we solve an algebraic equation numerically by Newton's method, we have to iterate a map similar to (3.29). If the equation has several solutions, an initial value for the iteration will be attracted to one of the solutions. The boundary of the set of points which finally converge to one of the solutions becomes a fractal. Two initial points which are arbitrarily close can approach distinct solutions, if they start close to this boundary.

Another simple method to construct fractals is provided by *contraction maps*. A contraction map is a mapping that shortens the distance between any two points. It is trivial that the invariant set of a single contraction map is a point. However, for two or more contraction maps the invariant set is the set X which satisfies

Figure 3.11 Julia set on the complex plane.

Figure 3.12 Mandelbrot set [119].

$$x = f_1(x) \cup f_2(x) \cup \ldots \cup f_n(x), \tag{3.30}$$

which is a fractal [59].

For example, in the case $n = 2$, the following maps produce the Cantor set in the interval [0, 1].

$$f_1(x) = \frac{x}{3}, \quad f_2(x) = \frac{2 + x}{3}. \tag{3.31}$$

In the complex plane we have the Koch curve if the mappings are

$$f_1(z) = \alpha\,\bar{z}, \quad f_2(z) = (1 - \alpha)\bar{z} + \alpha,$$
$$\alpha = \frac{1}{2} + \frac{\sqrt{3}}{6}\,\mathrm{i}, \tag{3.32}$$

where $\bar{z}$ denotes the complex conjugate of z. Figure 3.10 can be obtained by the following maps on $\mathbb{R}^2$:

$$f_1\begin{pmatrix} x \\ y \end{pmatrix} = \frac{1}{2}\begin{pmatrix} x \\ y \end{pmatrix}, \quad f_2\begin{pmatrix} x \\ y \end{pmatrix} = \frac{1}{2}\begin{pmatrix} 2 - x \\ 1 + y \end{pmatrix}. \tag{3.31'}$$

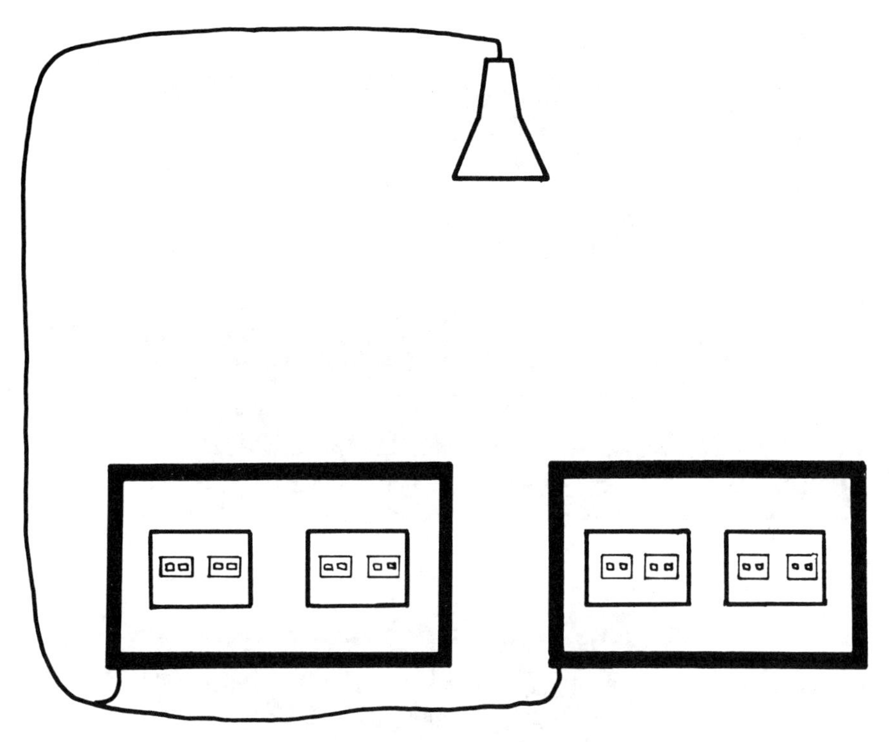

Figure 3.13 Cantor set produced by two televisions and one camera.

Thus all regular (non-random) fractals can be expressed in this formalism, which because its simplicity is expected to become more important in future.

Applying this idea, we can produce the Cantor set on a television screen with two television sets and a video camera. The method is to arrange the television sets in front of the camera as shown in Figure 3.13. Then on each screen we find two images of televisions in which we can further find two smaller images, and so on *ad infinitum*. The limit is the Cantor set.

3.3 **Random clusters**

3.3.1 *Percolation*

The percolation problem introduced in Section 2.4 can easily be simulated on a computer. Consider a 2- or 3-dimensional lattice and distribute points randomly on it with probability p. If neighbouring sites are occupied by points, they are regarded as connected. By changing the probability, p, of the occupation of sites we can estimate the critical probability p_c and fractal dimension of clusters (see Fig. 3.14).

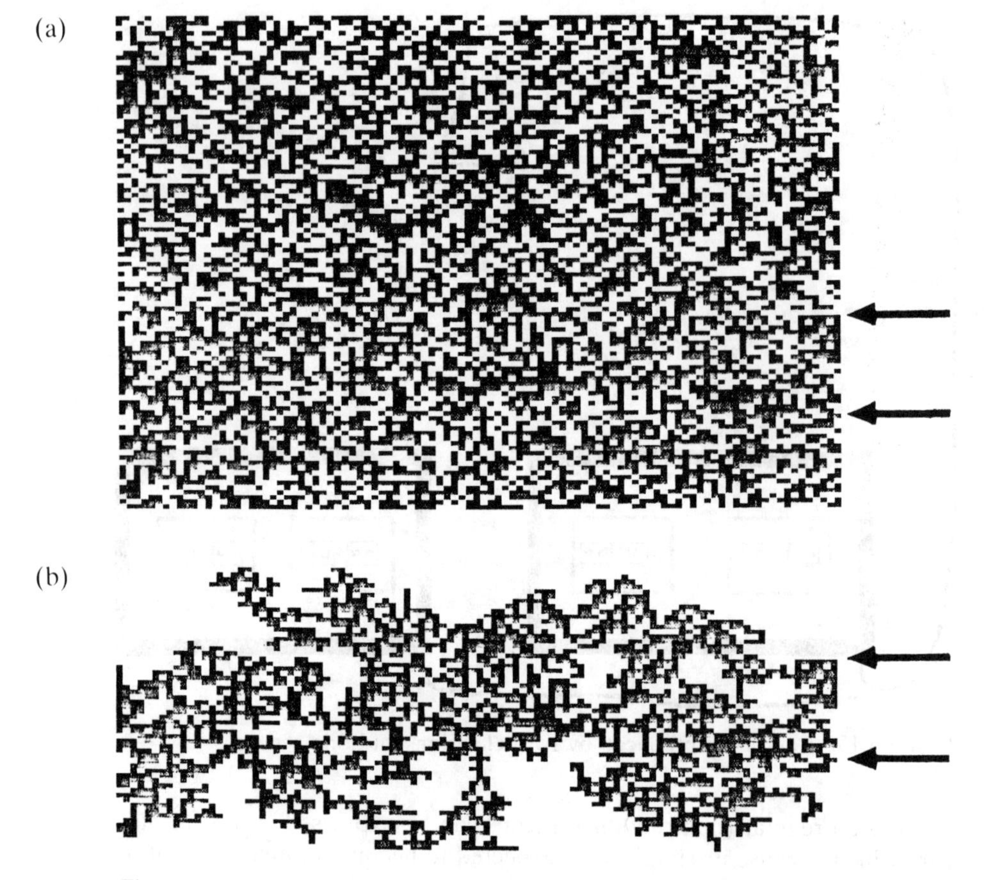

Figure 3.14 (a) Simulation of percolation; (b) the largest cluster.

The fractal dimension of clusters is calculated in the following way. We define the mean radius of clusters of size s as

$$R_s \equiv \left\langle \left(\sum_{i=1}^{s} \frac{r_i^2}{s} \right)^{1/2} \right\rangle, \tag{3.33}$$

where r_i denotes the distance between the centre of mass and the ith point, and $\langle \cdot \rangle$ indicates the average over all s-clusters. When R_s is proportional to a power of s, the clusters are statistical fractals with dimension D which satisfies the relation

$$R_s \propto S^{1/D}. \tag{3.34}$$

The results of simulations show that (3.34) holds at $p = p_c$ and the fractal dimensions are estimated as 1.9 (2-dimensional lattice) and 2.5 (3-dimensional lattice) [60]. This value in the 2-dimensional case agrees with the experimental value.

The critical point p_c is known to depend on the type of underlying lattice. On a square lattice $p_c = 0.59$, on a honeycomb lattice $p_c = 0.70$, and on a triangular lattice $p_c = 0.50$.

However, the fractal dimension and other critical indices are anticipated to be universal and independent of the underlying lattice.

3.3.2 *Clusters in spin systems*

The best-known model of magnetic material is the *Ising model*. In this model, spins which can take only the value +1 or −1 are arranged on a lattice. The total energy (or Hamiltonian), E, of the system is given by the equation

$$E = -J \sum\sum S_i S_j - H \sum_i S_i, \quad S_i = \pm 1. \tag{3.35}$$

Here, $\Sigma\Sigma$ denotes summation over nearest neighbour sites, J is the coupling constant and H is the external field. In thermal equilibrium, the probability of occurrence of the state with total energy E is given by

$$W \propto \exp(-E/k_B T), \tag{3.36}$$

where k_B is Boltzmann's constant and T denotes temperature.

A numerical simulation is performed as follows. First, specify an appropriate initial state, which may be random or uniform. Then, choose one spin at random and calculate the change of total energy of the system assuming that the spin is reversed. Change the sign of the spin according to the probability calculated from (3.36). Choose another spin at random and repeat the same process. After a large number of repetitions, thermal equilibrium is obtained.

In both 2- and 3-dimensional space, the Ising model is known to show a phase transition at a critical temperature, T_c. For $T < T_c$, symmetry is spontaneously broken and most spins take the same value, which indicates that the system is ferromagnetic. On the other hand when $T > T_c$, each spin takes the value +1 or −1 nearly independent of neighbouring spins and the average of spin vanishes, which shows that the system is demagnetised. At the critical point $T = T_c$, the characteristic size of clusters of the same spin diverges and distribution of the clusters becomes fractal. The fractal dimensions of the clusters are estimated to be 1.88 in 2-dimensional space and 2.43 in 3 dimensions [61].

3.4 **Electric breakdown and fracture**

There are great similarities between electric breakdown and fracture phenomena – in fact the shape of a lightning discharge resembles that of a crack in rock. Perhaps electric breakdown should be thought of as a kind of general fracture phenomenon.

There are two ways to simulate such fracture phenomena on a computer: one being stochastic and the other deterministic.

3.4.1 *A stochastic model of electric breakdown*

We assume that the electric breakdown between a pair of electrodes spreads stochastically with probability proportional to the local electric field. The growth procedure is given as follows:

Let ϕ denote the electric potential which takes values 1 or 0 at the electrodes. Solve the discrete version of the Laplace equation

$$\Delta\phi = \frac{\partial^2}{\partial x^2}\phi + \frac{\partial^2}{\partial y^2}\phi = 0, \tag{3.37}$$

on a lattice space with the given boundary condition. Choose a site at random from the neighbouring sites of the electrode with $\phi = 1$, with probability proportional to the gradient of ϕ. The bond connecting this site to the electrode is then regarded as broken and the value of ϕ becomes 1 at the site. Solve the Laplace equation with the new boundary condition and again choose at random a neighbouring site of the electrode and its attached cluster, with probability proportional to the gradient of ϕ. Repeat the same procedure again and again.

The cluster of sites forms the shape of a lightning discharge. It has been numerically and mathematically confirmed that the shape is identical to that of DLA (Section 3.1) [62].

This model has been extended to the more general η-model, in which we choose a broken site with probability proportional to $|\nabla\phi|^\eta$. The basic model is the special case $\eta = 1$. When $\eta > 1$ the difference of gradients, $|\nabla\phi|$, is enhanced, hence sites near a sharp tip of the cluster are more likely to be broken and we have a cluster with smaller fractal dimension. On the other hand when $\eta < 1$ we obtain clusters with larger dimension. The fractal dimension of a cluster which grows from a point-like electrode in d-dimensional space is approximately given by the following [63]:

$$D = \frac{d^2 + \eta}{d + \eta}. \tag{3.38}$$

Examples are shown in Figure 3.15.

3.4.2 *A deterministic model of fracture*

The other method of simulating fracture phenomena is based on a deterministic model of fracture [64].

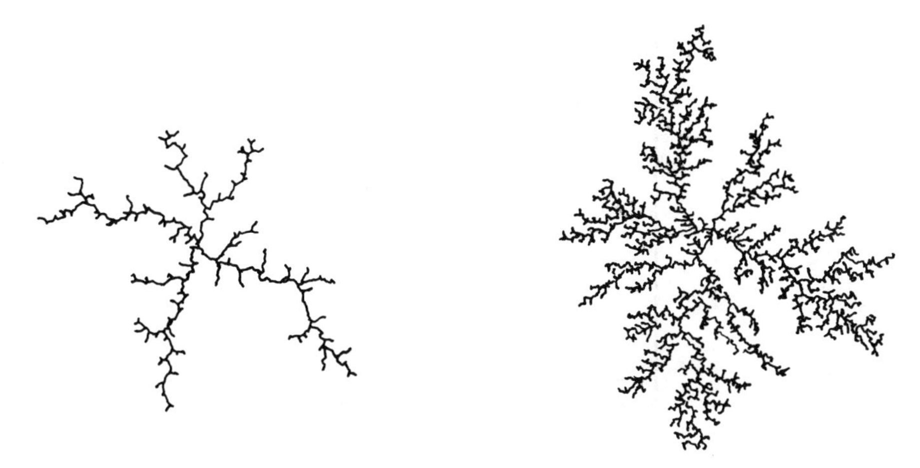

Figure 3.15 Clusters created by η-model: (a) $\eta = 1/2$; (b) $\eta = 2$.

In order to make a model of fracture, let us first consider a simple brittle fracture. Assume that one end of a thin brittle stick is fixed while the other end is free. If the displacement of the free end, $\delta\phi$, is less than a critical value, ϕ_c, then its modulus of rigidity is a constant, σ. Once $\delta\phi$ exceeds ϕ_c, the stick breaks and σ is suddenly reduced to a very small value (or zero). This nonlinear response is illustrated in Figure 3.16(a).

Next, we consider a planar square grid of brittle sticks. If we assume the displacements at the lattice points to be all perpendicular to the plane (the anti-plane shear problem), then the equation for the balance of forces becomes

$$\sum_{k=1}^{4} \sigma_k\,(i,\,j)\,\{\phi_k(i,\,j) - \phi(i,\,j)\} = 0, \tag{3.39}$$

where $\phi\,(i, j)$ denotes the displacement at (i, j), the suffix k specifies four directions, and $\sigma_k\,(i, j)$ denotes the rigidity of the corresponding stick.

Time evolution of this system is carried out by the following procedure:

(1) Give $\{\sigma\}$ and a boundary condition of $\{\phi\}$.
(2) Solve $\{\phi\}$ by (3.39).
(3) Check every unbroken stick. If the breakdown condition $|\phi_k\,(i, j) - \phi\,(i, j)| > \phi_c$ is satisfied, then the stick is regarded as broken and its rigidity is replaced by a very small one.
(4) Stop when no stick has been broken in the preceding procedure. Otherwise, go back to procedure (2) and continue the routine.

In this procedure we deterministically evolve the system by solving (3.39) and checking the breakdown condition, repeatedly.

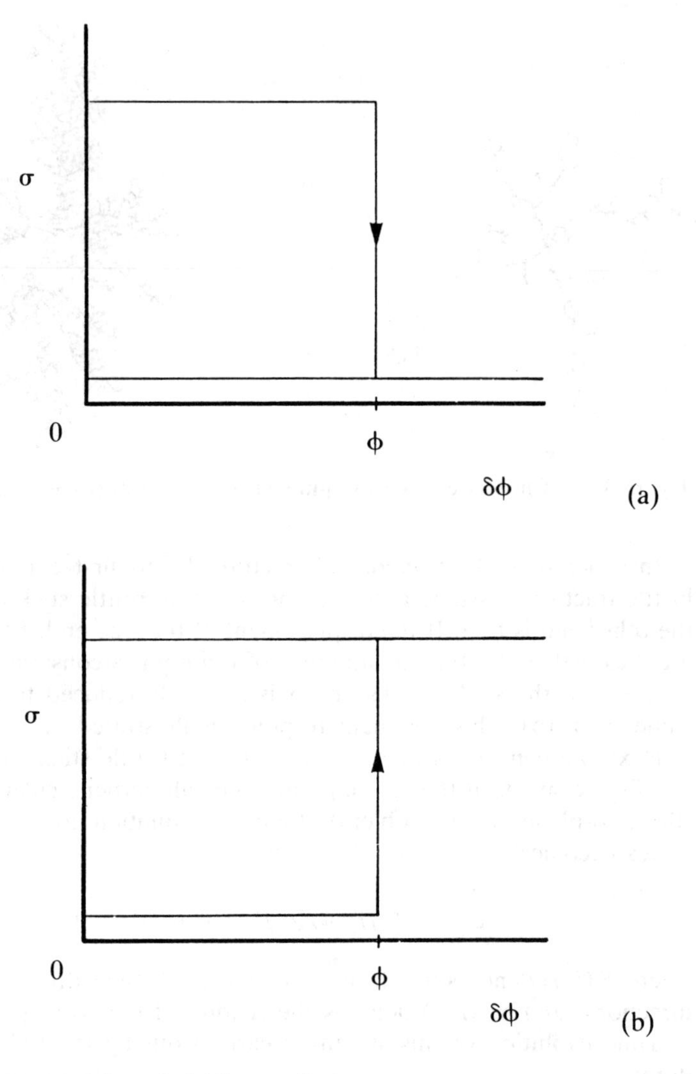

Figure 3.16 (a) Modulus of rigidity v. displacement in a brittle fracture; (b) conductivity v. voltage in electric breakdown.

In Figure 3.17(a) is shown an example of evolution on a 10×10 grid. Here, the initial values of $\{\phi\}$ have random fluctuations, and as boundary condition we assume that the top edge of the square brittle plate is pulled up while the bottom edge is fixed. We can find from this figure that the breakdown starts from a single stick and grows horizontally accompanying a bifurcation of the plate. The growth stops when the cracks form a percolation cluster. It should be noted that the randomness of the resulting pattern comes entirely from the initial randomness.

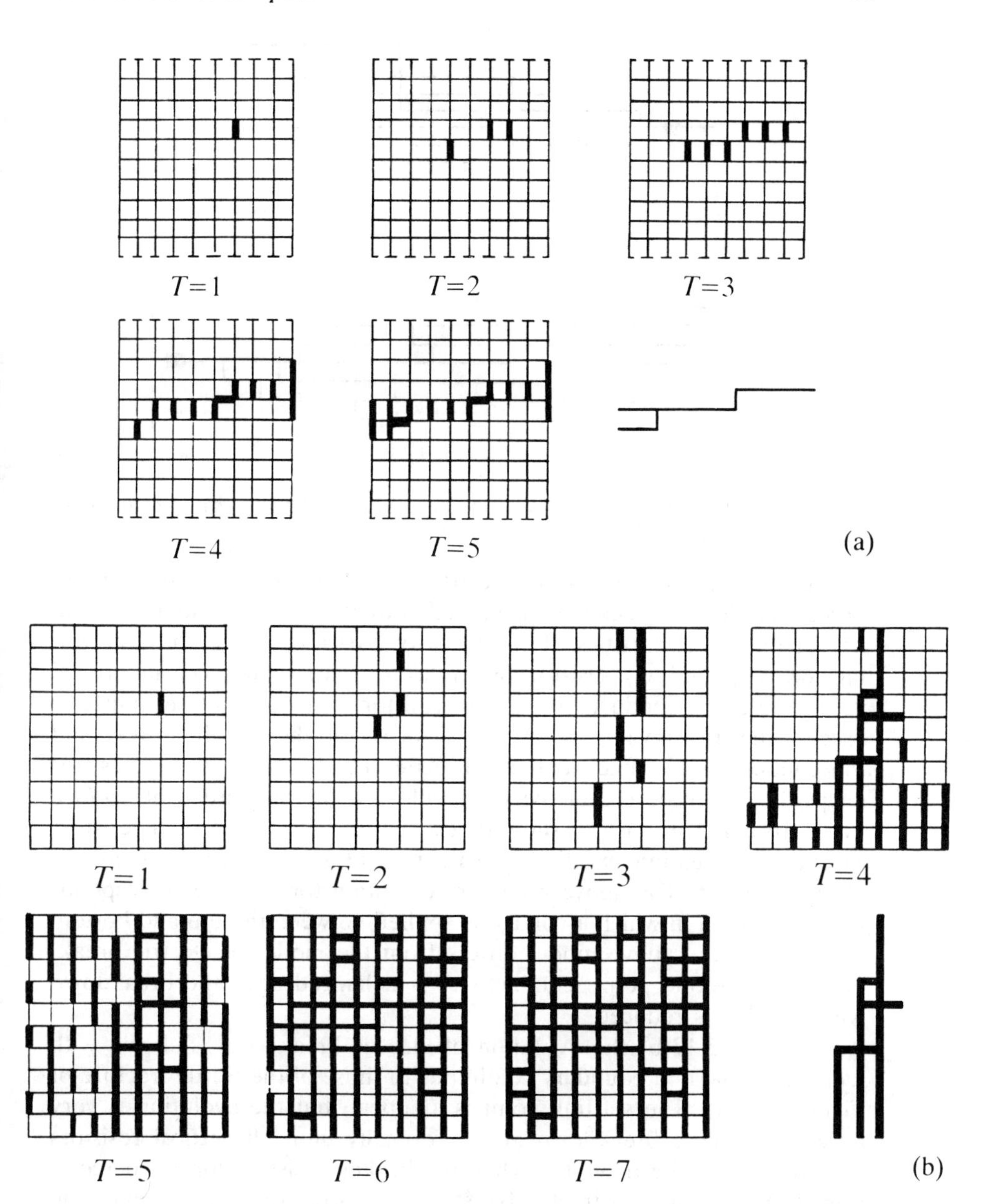

Figure 3.17 (a) An example of evolution of brittle fracture; (b) an example of evolution of electric breakdown. In (a) and (b) the last figures show the percolation cluster.

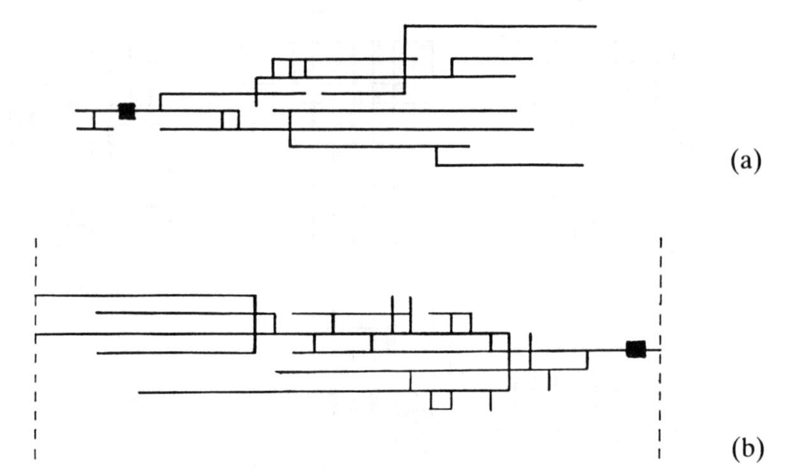

Figure 3.18 (a) An example of a crack on a 32 × 32 lattice; (b) an example of lightning on a 32 × 32 lattice.

We can apply this model to electric breakdown [65]. The elementary process of electric breakdown is modelled by the following nonlinear and irreversible characteristics of conductivity. Assume that a constant voltage $\delta\phi$ is applied to a resistor. When $\delta\phi$ is smaller than a certain critical voltage ϕ_c, the conductivity of the resistor is a constant, σ. Once it exceeds the critical value, then the resistor is considered to be broken and the conductivity is much enhanced. After the breakdown, the resistor keeps the enhanced conductivity regardless of the magnitude of $\delta\phi$ (see Fig. 3.16(b)). It is easy to show that a square network of such resistors satisfies an equation exactly equivalent to (3.39) in order to conserve electric current. The above boundary condition for fracture corresponds to the situation in which a voltage is applied between the top and bottom edges. Thus the above model differs from the model of electric breakdown in only one point: conductivity is enhanced after the breakdown while rigidity is reduced.

In Figure 3.17(b) the evolution of electric breakdown is shown with identical initial and boundary conditions to those of the brittle fracture of Figure 3.17(a). The starting point is identical, but the evolution is very different: electric breakdown grows vertically until all vertical resistors are broken. At the time step when the broken resistors form a percolation cluster, we have a fractal-like structure corresponding to a lightning discharge.

The fractal properties of the percolation clusters in both cases are confirmed by performing larger simulations. In Figure 3.18 clusters on a 32 × 32 lattice are shown. They have dendritic structures and look similar to those in Figure 3.17. The fractal dimensions for these clusters are estimated to be about 1.6 in both cases.

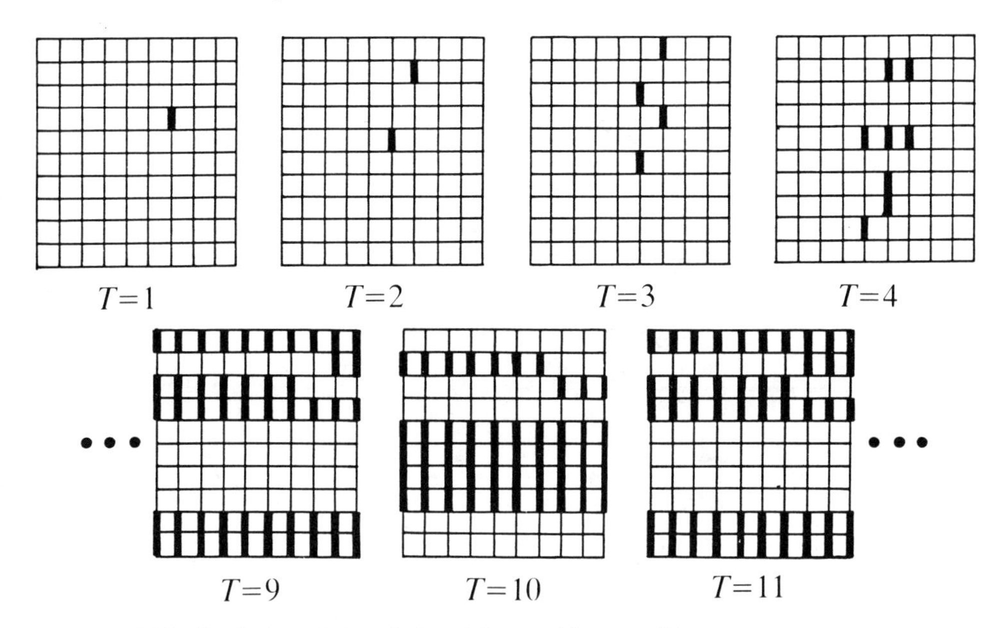

Figure 3.19 Evolution of electric breakdown with reversible response.

In order to make a fractal structure, the irreversible response is essential in electric breakdown. If we replace the response of the resistors by a nonlinear and reversible one, the result becomes very different. In such a case we have oscillations as shown in Figure 3.19, and no fractal structure appears. On the other hand, in the fracture problem, the behaviour does not change under such a change of rigidity.

In summarising this section, we point out the relation between the stochastic model and the deterministic model. The equation we have solved in the deterministic model (Equation (3.39)) becomes the following elliptic partial differential equation in the continuum limit:

$$\nabla \cdot (\sigma \nabla \phi) = 0. \tag{3.40}$$

The Laplace equation used in the stochastic model is obtained in the special case when σ is a constant. In the stochastic model there is no explicit spatial randomness but a spatial fluctuation is implicit in the stochastic manner of choosing growth sites. We may consider σ in (3.40) as a random function of time. Thus, considering the two models, we can say that the simple equation (3.40), which describes conservation of flux $\sigma\nabla\phi$, plays an essential role in the pattern formation of electric breakdown and brittle fracture. In other words, we may say that the origin of such dendritic fractal structures is in the conservation law of flux. (Note that DLA and viscous fingers are also included in this category. In DLA,

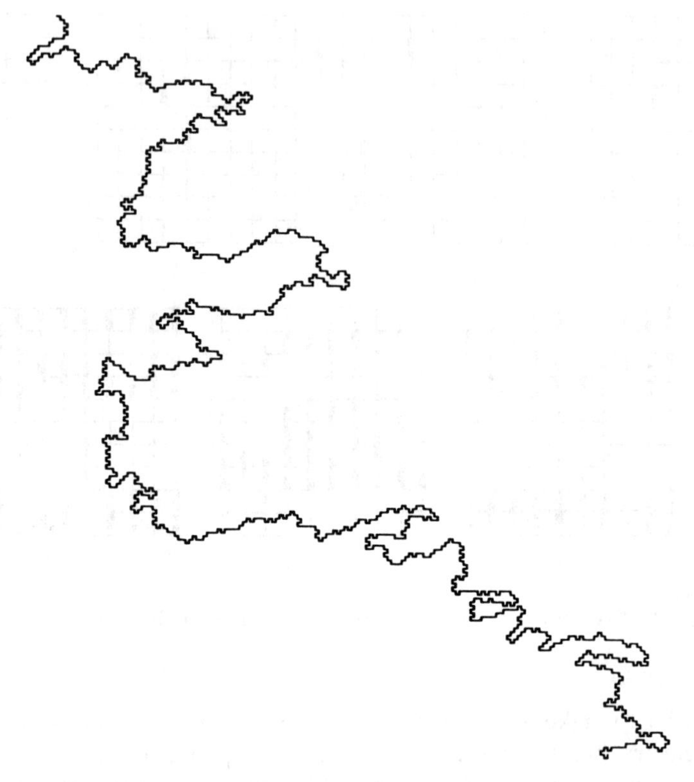

Figure 3.20 An example of self-avoiding random walk.

ϕ denotes the probability of the existence of a randomly walking particle, and for viscous fingers, it corresponds to the pressure.)

3.5 Self-avoiding random walks

A *self-avoiding random walk* is a random walk that never intersects its own trajectory. Though this condition is very simple, theoretical treatment becomes extremely difficult, since the whole past trajectory affects the present motion. In a computer it is easy to remember the whole trajectory and to avoid intersection; with a simple program we can perform a large and accurate simulation if the memory capacity is sufficiently large. An example of a self-avoiding random walk is shown in Figure 3.20; see Section 3.7 for the program listing.[2]

Self-avoiding random walks are considered as a model of a polymer. Thread-like polymers in solution are self-avoidingly entangled by thermal fluctuation. The fractal dimension of self-avoiding random walks in 3-dimensional space is obtained approximately as 5/3 which coincides with the experimental values for polymers given in Section 2.4. This value 5/3 can also be deduced theoretically by a dimensional analysis (see Section 5.3).

3.6 Cellular automata

Is it possible to produce complicated structures by a simple rule? Considering the fact that any living creature is formed from a finite amount of DNA, the idea of producing complicated structure by simple rules seems promising. The numerical models called *cellular automata* are studied in order to clarify this problem.

A cellular automaton has the following five properties:

(1) It is defined on a discrete lattice.
(2) Time evolution is discrete.
(3) The number of states at each site is finite.
(4) The rule of evolution is deterministic
(5) The evolution rule is governed by the state of neighbouring sites.

However, note that non-deterministic rules have also been studied.)

Here is the simplest non-trivial example. Let $a_i(n)$ denote the state of the ith site on a 1-dimensional lattice at time step n. The value taken by $a_i(n)$ is either 1 or 0. The evolution rule is given as

$$a_i(n) = a_{i-1}(n-1) + a_i(n-1) \mod 2. \tag{3.41}$$

Figure 3.21 shows the time evolution of this automaton with the initial conditions

$$\begin{aligned} a_0(0) &= 1, \\ a_i(0) &= 0, \quad i \neq 0. \end{aligned} \tag{3.42}$$

In the figure, a dot represents $a_i(n) = 1$. This pattern is nothing but Sierpinski's gasket in discrete space-time. Remarkably, such a simple rule as (3.41) together with the simple initial condition (3.42) produces a complicated fractal structure.

The properties of this system with periodic boundary condition has been analysed in detail in [66]. The final state of this system must be a fixed point or a limit cycle since the total number of states is finite (2^M when the number of sites is M). An interesting problem is how the number and the length of limit cycles change as functions of M. In the case $M = 2^k$, beginning with any initial conditions, the system will fall into a null state (all sites have $a_i = 0$). When M is a prime number the primitive root of which is 2, there is one limit cycle of length $2^{M-1} - 1$. In other cases M has multiple limit cycles. There, initial small perturbations sometimes cause large differences in final states.

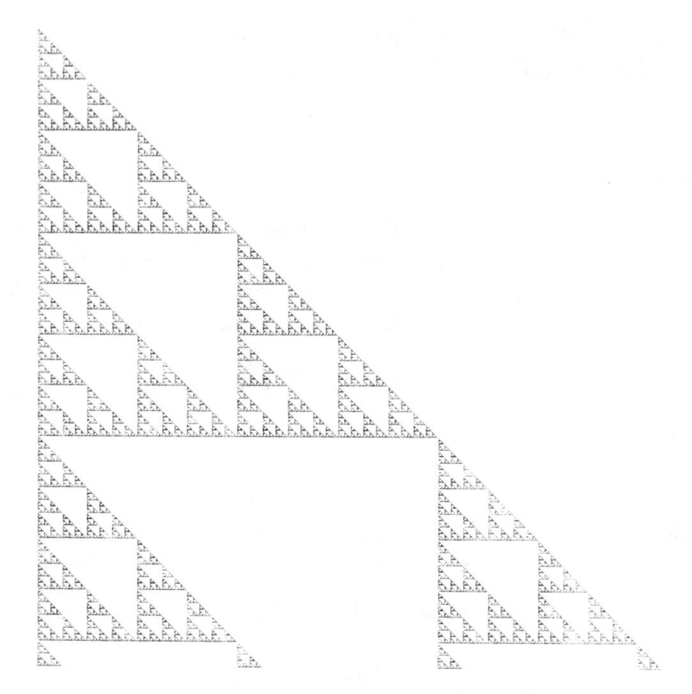

Figure 3.21 A space–time pattern of a cellular automaton.

A system with multiple limit cycles shows a complicated behaviour like a system with a strange attractor. However, the complexities of cellular automata and chaotic systems are expected to be basically different, since the sources of complexity are different. The source of the complexity of cellular automata is the large number of degrees of freedom while that of chaotic systems is the nonlinearity.

Recently, cellular automata have been quantitatively classified into four types by their asymptotic behaviours following a random start [67].

Type 1. The system is attracted to a stationary state.
Type 2. It is a attracted to a limit cycle and shows a periodic motion.
Type 3. The system remains disordered for all time and throughout space. Correlations have short range.
Type 4. Localised disordered regions survive, and the final state is not predictable even on a coarse-grained level.

The above mentioned rule which produces the fractal pattern belongs to Type 3. In Type 4, no abbreviation of the calculation is supposed to exist, hence we have to evolve the whole system step by step in order to know its asymptotic behaviour. In a system of infinite size, we need infinite time to check the final state. Therefore, Type 4 is unpredictable.

The above classification cannot be made directly on the basis of the evolution rule, but in general numerical simulation is necessary. The cases that produce fractal patterns are exceptions, in which theoretical analysis by means of the renormalisation group is possible. The most difficult case is Type 4 which may be regarded as more complicated than fractals.

3.7 Program listings

This section provides listings of eleven Basic programs which have been used to draw some of the figures in this book. The version of Basic is N88BASIC for NEC PC9800 personal computer. All the programs are short and may easily be translated into other computer languages. Running these programs will help the reader become familiar with fractals.

3.7.1 *The Koch curve*

There are many methods of drawing the Koch curve (see Fig. 1.1). Here we use contraction maps as described in Section 3.2.3.

```
100 '
110 ' VON KOCH CURVE
120 '
130 N=12:PI=3.14159
140 DIM X(2^(N+1)-2),Y(2^(N+1)-2)
150 SCREEN 2,0:CLS 3
160 WINDOW(0,-2/3)-(1,0)
170 VIEW(0,0)-(599,399)
180 '
190 A=SQR(1/3)*COS(PI/6)
200 B=SQR(1/3)*SIN(PI/6)
210 A1=A:A2=B:A3=B:A4=-A
220 B1=A:B2=-B:B3=-B:B4=-A
230 '
240 X(0)=0:Y(0)=0
250 FOR M=1 TO N
260  L2=2^(M-1)-1:L1=L2*2+1:L3=L1*2
270  FOR K=0 TO L2
280    XX=X(L2+K):YY=Y(L2+K)
290    X(L1+K)=A1*XX+A2*YY
300    Y(L1+K)=A3*XX+A4*YY
310    X(L3-K)=B1*XX+B2*YY+1-B1
320    Y(L3-K)=B3*XX+B4*YY-B3
330    PSET(X(L1+K),-Y(L1+K))
340    PSET(X(L3-K),-Y(L3-K))
350  NEXT K
360 NEXT M
```

Figure 3.22 A fractal branch.

By changing the contraction maps we can make many fractal shapes. For example, a self-similar branch (Fig. 3.22) is obtained by replacing the parameters in line 220 as shown in the following list.

```
100 '
110 ' FRACTAL BRANCH
120 '

160 WINDOW(0,-1/3)-(1,1/3)

220 C=2/3: B1=C:B2=0:B3=0:B4=-C
```

3.7.2 *Lévy dust*

See Figure 1.15. The fractal dimension of the dust changes if we change D in line 130.

```
100 '
110 ' LEVY FLIGHT 2-D
120 '
130 D=1.5
140 DD=-1/D:P2=3.14159*2
150 XL=100+10^(-DD*3.5):YL=XL
160 '
170 RANDOMIZE
180 '
190 SCREEN 3:CLS 3
200 WINDOW(-XL,-YL)-(XL,YL)
210 VIEW(0,0)-(399,399)
220 LINE(-XL,-YL)-(XL,YL),7,B
230 '
240 N=1:LOCATE 70,0:PRINT"N=";N
250 X=0:Y=0:PSET(X,Y)
260 '
270 *MAIN
280 Z=(1-RND)^DD:W=RND*P2
290 XX=X+Z*COS(W):YY=Y+Z*SIN(W)
300 X=XX:Y=YY:PSET(X,Y),1+(N/100 MOD 7)
310 N=N+1:LOCATE 70,0:PRINT"N=";N
320 GOTO *MAIN
```

3.7.3 *Diffusion-limited aggregation*

Much computer time is needed to make a large cluster, example, Figure 3.2 took several hours.

```
100 '
110 '           Aggregation on 2D Lattice
120 '
130 CLS 3
140 SCREEN 2
150 P=320 : Q=200                 ' Location of the seed
160 R0=5                          ' Initial value of R0
170 PSET(P,Q)
180 '
190 *MAIN
```

```
200 R=R0 * 2                      ' Particles appear at R
210 RMAX=R0 * 3                   ' Limit of moving area
220 RX=INT((2*R+1)*RND)-R
230 RV=R-ABS(RX)
240 RY=RV*SGN(RND-.5)
250 X=RX+P : Y=RY+Q
260 '
270 *LOOP
280 XB=X : YB=Y
290 DISTR = ABS(X-P)+ABS(Y-Q)
300 IF  POINT(X,Y-1)=1  OR  POINT(X,Y+1)=1  OR
    POINT(X-1,Y)=1  OR  POINT(X+1,Y)=1  THEN  *AGGR
310 '
320 IF DISTR > RMAX THEN PRESET(X,Y) : GOTO *MAIN
330 TWD(1)=0 : TWD(2)=0
340 TWD( INT(2*RND)+1 ) = SGN(RND-.5)
350 X=X+TWD(1) : Y=Y+TWD(2)
360 PRESET(XB,YB) : PSET(X,Y)
370 GOTO *LOOP
380 '
390 *AGGR
400 PSET(X,Y)
410 IF DISTR > R0 THEN R0=DISTR
420 GOTO *MAIN
```

3.7.4 *The Lorenz attractor*

See Figure 3.4.

```
100 '
110 ' LORENZ MODEL
120 '
130 X=10:Y=12:Z=15:R=50:DT=.001:T=0
140 SCREEN 2.0:CLS 3
150 DIM V(1.3).W(1.3).U(1)
160 FOR I=0 TO 1
170 FOR J=0 TO 3:READ W(I.J):NEXT J
180 FOR J=0 TO 3:READ V(I.J):NEXT J
190 WINDOW(W(I.0).W(I.1))-(W(I.2).W(I.3))
200 VIEW(V(I.0).V(I.1))-(V(I.2).V(I.3))
210 LINE(-200.0)-(200.0):LINE(0.-200)-(0.200)
220 NEXT I
230 LOCATE 37.34:PRINT"X".:LOCATE 18.3:PRINT"Z".
240 LOCATE 78.34:PRINT"Y".:LOCATE 60.2:PRINT"Z".
```

```
250 '
260 *MAIN
270 T=T+DT:LOCATE 40,0:PRINT"T=";T
280 U(0)=X:U(1)=Y
290 FOR I=0 TO 1
300 WINDOW(W(I,0),W(I,1))-(W(I,2),W(I,3))
310 VIEW(V(I,0),V(I,1))-(V(I,2),V(I,3))
320 PSET(U(I),-Z)
330 NEXT I
340 XX=X+(-10*(X-Y))*DT
350 YY=Y+(-X*Z+R*X-Y)*DT
360 ZZ=Z+(X*Y-2.66667*Z)*DT
370 X=XX:Y=YY:Z=ZZ
380 GOTO *MAIN
390 DATA -30,-100,40,5,0,50,299,349
400 DATA -40,-100,50,5,330,50,629,349
```

3.7.5 *The attractor of the Hénon map*

It is better to compute in double precision to produce a more magnified structure than Figure 3.5.

```
100 '
110 ' HENON MAP
120 '
130 DIM C(3),D(3),W(3,3),V(3,3)
140 A=1.4 :B=.3 :XC=.83:YC=.15:VC=99.5
150 D(0)=2.5:D(1)=.4:D(2)=.08:D(3)=.0125
160 C(0)=XC:C(1)=YC:C(2)=XC:C(3)=YC
170 FOR I=0 TO 3:FOR J=0 TO 3
180 W(I,J)=C(J)+(2*INT(J/2)-1)*D(I)
190 READ V(I,J)
200 NEXT J:NEXT I
210 '
220 SCREEN 2,0:CLS 3
230 FOR I=0 TO 2
240 WINDOW(W(I,0),W(I,1))-(W(I,2),W(I,3))
250 VIEW(V(I,0),V(I,1))-(V(I,2),V(I,3))
260 LINE(W(I,0),W(I,1))-(W(I,2),W(I,3)),1,B
270 LINE(W(I+1,0),W(I+1,1))-(W(I+1,2),W(I+1,3)),1,B
280 NEXT I
290 WINDOW(W(3,0),W(3,1))-(W(3,2),W(3,3))
300 VIEW(V(3,0),V(3,1))-(V(3,2),V(3,3))
310 LINE(W(3,0),W(3,1))-(W(3,2),W(3,3)),1,B
```

```
320 '
330 X=1:Y=1
340 FOR K=0 TO 20
350 XX=1+Y-A*X*X:YY=B*X:X=XX:Y=YY
360 NEXT K
370 '
380 *MAIN
390 XX=1+Y-A*X*X:YY=B*X
400 FOR I=0 TO 3
410 WINDOW(W(I,0),W(I,1))-(W(I,2),W(I,3))
420 VIEW(V(I,0),V(I,1))-(V(I,2),V(I,3))
430 PSET(XX,YY)
440 NEXT I
450 X=XX:Y=YY:K=K+1
460 LOCATE 65,1 :PRINT K
470 GOTO *MAIN
480 DATA 0,0,199,199,200,0,399,199
490 DATA 0,200,199,399,200,200,399,399
```

3.7.6 *The Julia set*

See Figure 3.11. Changing the parameters AR and AI in line 130, can produce many strange shapes. Beautiful pictures can be obtained if the point (ZR, ZI) is coloured at line 280. The Mandelbrot set can be made by simple modification of the program.

```
100 '
110 '     JULIA SET   (F(Z)=A*Z*(1-Z))
120 '
130 AR=3.3:AI=0
140 NX=499:NY=399:R=4:TM=50
150 XL=0!:XU=1!:YL=-.25:YU=.25
160 SCREEN 2:CLS 3
170 WINDOW (XL,-YU)-(XU,-YL):VIEW (0,0)-(NX,NY)
180 DX=(XU-XL)/NX:DY=(YU-YL)/NY
190 '
200 FOR I=XL+DX TO XU-DX STEP DX
210  FOR J=YL+YX TO YU-YX STEP DY
220  ZR=I:ZI=J:T=0
230  ZR2=ZR*ZR:ZI2=ZI*ZI
240  *LOOP
250  ZZR=ZR-ZR2+ZI2:ZZI=ZI*(1-2*ZR)
260  ZR=AR*ZZR-AI*ZZI:ZI=AR*ZZI+AI*ZZR
270  ZR2=ZR*ZR:ZI2=ZI*ZI
```

```
280  IF ZR2+ZI2>R GOTO  *GO
290  T=T+1:IF T<TM GOTO *LOOP
300  PSET (I,-J)
310  *GO
320 NEXT:NEXT
```

3.7.7 *Percolation*

See Figure 3.14. The main part of the program in which points are plotted random is very short. Most of the program is used for specifying connected clusters. (The function POINT (X, Y) gives the colour at the point (X, Y) on screen, for example, black is 0, red is 1.)

```
100 '
110 '   Percolation
120 '
130 SCREEN 3,0:DEFINT B-Z:RANDOMIZE
140 INPUT"probability p=";A :CLS 3
150 NX=20:NY=20:NX1=NX+1:NY1=NY+1
160 XX=10:YY=10:X1=XX-1:Y1=YY-1
170 DIM P(NX1,NY1),COUNT(3000):KP=1
180 '       cluster creation
190 FOR X=0 TO NX1:FOR Y=0 TO NY1
200  IF X=0 OR X=NX1 OR Y=0 OR Y=NY1 THEN P(X,Y)=-1:GOTO *NE
210  IF RND<A THEN GOTO *NE
220  LINE((X-1)*XX+1,((Y-1)*YY+1))-STEP(X1,Y1),7,BF:P(X,Y)=-1
230 *NE:NEXT:NEXT
240 LINE(0,0)-(NX*XX+1,NY*YY+1),7,B
250 '       paint routine
260 FOR X=1 TO NX:FOR Y=1 TO NY
270  IF P(X,Y)=0 THEN *PA ELSE *NE2
280  *PA:PAINT ((X-1)*XX+1,(Y-1)*YY+1),3,7
290  IF P(X-1,Y) AND P(X+1,Y) AND P(X,Y+1) AND P(X,Y-1) THEN *PB
300  FOR XP=X TO NX:FOR YP=1 TO NY
310   IF POINT((XP-1)*XX+1,(YP-1)*YY+1)=3 THEN P(XP,YP)=KP
320  NEXT:NEXT
330  *PB:PAINT((X-1)*XX+1,(Y-1)*YY+1),0,7
340  KP=KP+1
350 *NE2:NEXT:NEXT
360 '       count routine
370 FOR X=1 TO NX:FOR Y=1 TO NY
380  IF P(X,Y)<>-1 THEN COUNT(P(X,Y))=COUNT(P(X,Y))+1
390 NEXT:NEXT
400 FOR I=1 TO KP
410  IF COUNT(I)>MAX THEN MAX=COUNT(I):MAXP=I
420 NEXT
430 '       output max cluster
440 'SCREEN 3,0,1,17:CLS 2
450 FOR X=1 TO NX:FOR Y=1 TO NY
460  IF P(X,Y)<>MAXP THEN *NE3
470  LINE((X-1)*XX+1,(Y-1)*YY+1)-STEP(X1,Y1),2,BF
480 *NE3:NEXT:NEXT
```

3.7.8 *Self-avoiding random walk*

See Figure 3.2.

```
'
'    SELF-AVOIDING RANDOM WALK
'
K=2     'step length
RANDOMIZE
SCREEN 3 : CLS 3
TWD(INT(2*RND))=SGN(RND-.5)
X=320 : XB=X : PX(1)=TWD(0)  : X=X+K*PX(1)
Y=200 : YB=Y : PY(1)=TWD(1)  : Y=Y-K*PY(1)
LINE (X,Y)-(XB,YB)
*LOOP
X1=PX(1)  : Y1=PY(1)
PX(0)=-Y1 : PX(2)= Y1 : PX(3)= X1-Y1 : PX(4)= X1+Y1
PY(0)= X1 : PY(2)=-X1 : PY(3)= X1+Y1 : PY(4)=-X1+Y1
FOR I=0 TO 4 : P(I)=POINT(X+K*PX(I),Y-K*PY(I)):NEXT I
IF P(0) OR P(2) OR P(3) OR P(4) <> 0 THEN *OD
TWD=INT(3*RND) : GOTO *WALK
*OD
SGNR=SGN(ROT) : IF SGNR=0 THEN *MA
TWD=1-SGNR   : IF P(TWD)=1 THEN *MA
GOTO *WALK
*MA
I=0
FOR J=0 TO 2
  IF P(J)=0 THEN SP(I)=J : I=I+1
NEXT J
IF I=0 THEN *BACK
TWD=SP(INT(I*RND)) : GOTO *WALK
*WALK
ROT=ROT+TWD-1
PX(1)=PX(TWD) : XB=X : X=X+K*PX(1)
PY(1)=PY(TWD) : YB=Y : Y=Y+K*PY(1)
IF X<0 OR X>639 OR Y<0 OR Y>399 THEN END
LINE (X,Y)-(XB,YB)
GOTO *LOOP
*BACK
X1=PX(1) : PX(1)=-X1 : XB=X :X=X+K*PX(1)
Y1=PY(1) : PY(1)=-Y1 : YB=Y :Y=Y+K*PY(1)
LINE (X,Y)-(XB,YB),0 : PSET(XB,YB)
FOR I=-1 TO 1 STEP 2
```

```
500    IF POINT (X+I,Y)=1 THEN PX(1)=-I  : PY(1)= 0
510    IF POINT (X,Y+I)=1 THEN PX(1)= 0  : PY(1)= I
520 NEXT I
530 IF PX(1)=-Y1 AND PY(1)= X1 THEN ROT=ROT-1
540 IF PX(1)= Y1 AND PY(1)=-X1 THEN ROT=ROT+1
550 GOTO *LOOP
```

3.7.9 *Cellular automaton*

See Figure 3.19. By changing the initial condition and the evolution rule, many interesting patterns will be obtained.

```
100 '
110 ' AUTOMATON
120 '
130 NX=399:NT=399
140 DIM X(400),Y(400)
150 SCREEN 2,0:CLS 3
160 '
170 FOR I=0 TO NX
180 X(I)=0
190 NEXT I
200 X(1)=1
210 FOR I=0 TO NX
220 PSET(I,0),X(I)
230 NEXT I
240 '
250 FOR N=1 TO NT
260 Y(0)=(X(0)+X(NX)) MOD 2
270 PSET(0,N),Y(0)
280 FOR I=1 TO NX
290 Y(I)=(X(I)+X(I-1)) MOD 2
300 PSET(I,N),Y(I)
310 NEXT I
320 FOR I=0 TO NX
330 X(I)=Y(I)
340 NEXT I
350 NEXT N
```

3.7.10 *Fractional Brownian motion*

This is a generalised Brownian motion which will be introduced in Section 5.4. We can continuously change the fractal dimension of the graph by changing the parameter D in line 240. See Figure 3.23.

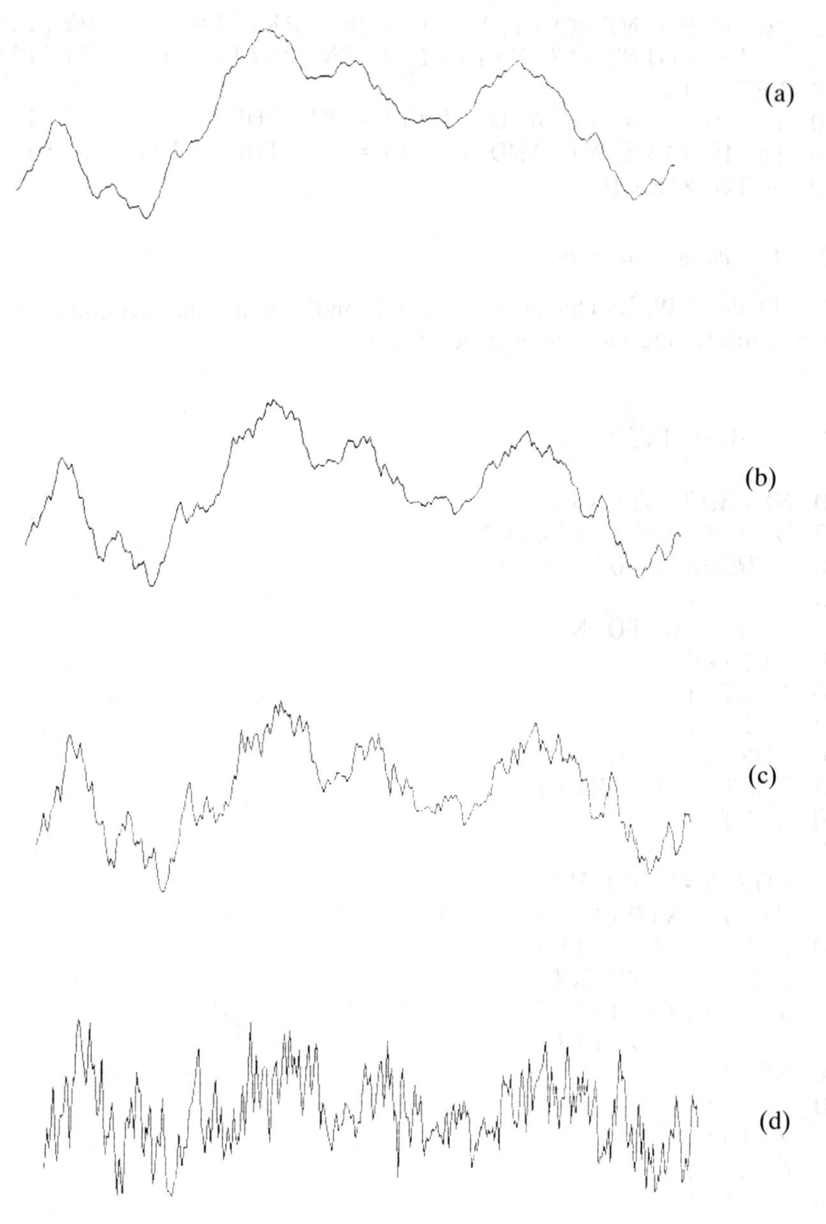

Figure 3.23 Examples of fractional derivative: (a) D = 1.0; (b) D = 1.25; (c) D = 1.5; (d) D = 2.0.

```
100 '
110 '   Fractional Derivative
120 '
130 SCREEN 2 : CONSOLE 0,25,0,0
140 KMAX=100:XMAX=300
150 RANDOMIZE
160 PI2=3.14159*2:EK=PI2/XMAX*.7
170 DIM ER(KMAX),EI(KMAX),KA(KMAX),F(XMAX)
180 FOR K=1 TO KMAX
190  GOSUB *GA : ER(K)=MO
200  GOSUB *GA : EI(K)=MO
210 NEXT K
220 *SA
230 CLS
240 INPUT "Fractal Dimension (1.0-2.0) ",D
250 A=5-2*D
260 FMAX=0 : FMIN=0
270 FOR K=1 TO KMAX : KA(K)=(EK*K)^(-A/2) : NEXT K
280 FOR X=0 TO XMAX
290  F(X)=0
300  FOR K=1 TO KMAX
310   KX=EK*K*X
320   F(X)=F(X)+KA(K)*(ER(K)*COS(KX)-EI(K)*SIN(KX))
330  NEXT K
340  IF FMAX<F(X) THEN FMAX=F(X)
350  IF FMIN>F(X) THEN FMIN=F(X)
360 NEXT X
370 '
380 WINDOW (0,FMIN*1.2)-(XMAX,FMAX*1.2)
390 VIEW (0,200)-(XMAX,299)
400 FB=F(0)
410 FOR X=1 TO XMAX
420  LINE (X-1,FB)-(X,F(X))
430  FB=F(X)
440 NEXT X
450 GOTO *SA
460 *GA
470 MO=0
480 FOR GI=1 TO 10
490  MO=MO+RND-.5
500 NEXT GI
510 RETURN
```

Notes

1 Analytical estimation of Lyapunov exponents is generally very difficult. Recently Doering *et al.* [110], have succeeded in obtaining exact values of Lyapunov exponents for a system governed by a nonlinear partial differential equation called the complex Ginzburg–Landau equation.

2 Technically there are several versions of self-avoiding random walks. The program in Section 3.7 is a so-called indefinitely growing self-avoiding walk. See Peliti [108] for detailed discussions.

4 Theoretical models of fractals

Solvable models are generally very simplified, but there are some great advantages in their theoretical treatments. One is that we can extract the essence of problems without being bothered by unknown elements such as noise or errors. Another merit is the generality of theoretical results: in experiments and numerical simulations, the size of the system and the number of realisations are limited while in theoretical treatments we are free from those restrictions.

4.1 Models of turbulence

We have already seen in Section 2.4 that turbulence has a fractal structure. Why is turbulence fractal instead of being homogeneous in space? Although no full answer is yet available, the following qualitative explanation may be indicative.

Let us regard turbulence as an ensemble of *vortex tubes*. A vortex tube is an imaginary cylindrical surface which connects points of equal vorticity. According to fluid dynamics, vortex tubes are stretched by their motion, that is, their length increaes with time. When the velocity field is turbulent, it is folded over and over again by turbulent flow. Provided the viscosity is negligibly small, the vortex tube has to be folded self-avoidingly, since according to Helmholtz's vortex theorem it cannot be torn.

The rate of energy dissipation $\varepsilon(x)$ mentioned in Section 2.4 is proportional to the square of the absolute value of vorticity. Therefore, the fact that the distribution of vorticity is fractal means that the distribution of energy dissipation is also fractal. Hence the energy dissipation region of turbulence has a fractal structure.

This idea can be developed quantitatively with the method of dimensional analysis [68]. Accordingly, the fractal dimension of turbulence, D, has been conjectured to be

$$D = \frac{2d + 7}{5}, \tag{4.1}$$

where d is the spatial dimension. If we set $d = 3$, the fractal dimension becomes $D = 2.6$, which is consistent with the experimental results, $2.5 < D < 2.8$.

The fractal distribution of vorticity is also called intermittency and is expected to affect several important quantities. An example is the energy spectrum. In Section 5.3. we will discuss the famous Kolmogorov 5/3 power law and its modification which takes into account the fractal effect. Here, we shall consider the velocity distribution by introducing a simple model [69].

Let us consider the situation where many point vortices are distributed in 3-dimensional Euclidean space. We denote their locations and vorticity by $\{\vec{r}_j\}$ and $\{\vec{w}_j\}$, respectively. Here a *point vortex* is an imaginary unit of the vortex field where vorticity is concentrated at a single point such as a delta function. If we assume that each point vortex generates a flow according to the Biot–Savart law, the velocity at $\vec{r}$ will be given by

$$\vec{u}(\vec{r}) = \sum_j \vec{u}_j, \tag{4.2}$$

$$\vec{u}_j \equiv -\frac{\vec{w}_j \times (\vec{r} - \vec{r}_j)}{4\pi \, |\vec{r} - \vec{r}_j|^3} \tag{4.3}$$

Assuming that vortices distribute randomly independent of each other, we obtain the distribution function of velocity $\vec{u}$ as

$$W(\vec{u}) = \prod_j \iiint d\vec{r}_j \, \tau_j(\vec{r}_j, \vec{w}_j) \cdot \delta(\vec{u} - \sum_j \vec{u}_j) \tag{4.4}$$

where $\tau_j(\vec{r}_j, \vec{w}_j)$ expresses the probability that the jth point vortex is at $\vec{r}_j$ with vorticity $\vec{w}_j$. If the point vortices distribute only on a D-dimensional fractal region F_D, then $\tau_j(\vec{r}_j, \vec{w}_j)$ is expressed as

$$\tau_j(\vec{r}_j, \vec{w}_j) \propto \begin{cases} \tau(\vec{w}_j), & \vec{r}_j \in F_D, \\ 0, & \vec{r}_j \notin F_D. \end{cases} \tag{4.5}$$

After a little calculation, we have $W(\vec{u})$ in the form

$$W(\vec{u}) = \frac{1}{8\pi^3} \iiint d\vec{\rho} \, \exp\left\{-i\vec{\rho} \cdot \vec{u} - \eta A\left(\frac{\vec{\rho}}{|\vec{\rho}|}\right) \cdot |\vec{\rho}|^{D/2}\right\}, \tag{4.6}$$

where η is a constant proportional to the density of vorticity on F_D, and $A(\vec{\rho}/|\vec{\rho}|)$ is given by

$$A\left(\frac{\vec{\rho}}{|\rho|}\right) \equiv \iiint d\vec{w}\ \tau(\vec{w}) \left| \vec{w} \times \frac{\vec{\rho}}{|\vec{\rho}|} \right|^{D/2}. \tag{4.7}$$

As we can see from this expression, its power exponent depends on the fractal dimension D. In the special case $D = 3$ (the case that vortices distribute uniformly), the distribution of velocity becomes the so-called Holtsmark distribution. The reader who is acquainted with the central limit theorem might think that this should be a Gaussian distribution, but it is actually far from Gaussian except in the neighbourhood of $|\vec{u}| = 0$, as shown in Figure 4.1.

The above model can be generalised in the following way [71]. Let a quantity $x(r)$ be produced by a linear transformation of white noise $w(r)$ with a Green function $G(r)$:

$$x(r) = \int_{-\infty}^{\infty} G(t - t')\ w(t')\ dt'. \tag{4.9}$$

In the case when $G(r)$ is proportional to a power of r such that

$$G(t) \propto |t|^{H-d}, \tag{4.10}$$

then (4.10) represents a fractional integral of order H in d-dimensional space (see Section 5.4 for discussion of fractional integral). The power spectrum of x becomes

$$S_x\ (k) \propto k^{3d-2D-2H-1}. \tag{4.11}$$

Here D denotes the fractal dimension of the region where the white noise exists. It has been shown that the distribution of x becomes a stable distribution with the characteristic exponent (see section 5.2)

$$\alpha = \frac{D}{d - H} \tag{4.12}$$

instead of $D/2$ in (4.6). The preceding result, (4.6), is included as the special case of $d = 3$ and $H = 1$. As seen from this example, a power law Green function generally results in a stable distribution having a long tail. This suggests that in a system with long-range interaction, such as gravity or electrical forces, we may find some fractal distributions.[1]

Thus, in the problem of turbulence, we have seen that fractal properties of the energy dissipation region, the power spectrum and velocity distributions are closely related to each other.

Now let us consider the fractal dimensions of clouds and *sumi-nagashi* (or marbling). Since both clouds and ink flow along the stream lines of a fluid, their fractal structures are expected to be formed by turbulent

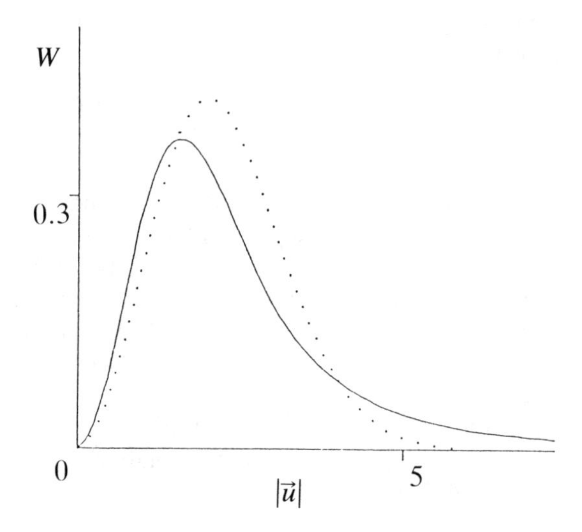

Figure 4.1 Gaussian distribution (dotted line) and Holtsmark distribution (solid line).

diffusion. A famous law about turbulent diffusion is that known as *Richardson's 4/3 power law*. According to this law, the distance R between two particles floating along turbulent flow increases with time so as to satisfy the following relation:

$$\frac{\mathrm{d}}{\mathrm{d}t}\langle R^2\rangle \propto R^{4/3} \tag{4.13}$$

If we pour indian ink into a turbulent flow, fine particles of the ink diffuse away from each other (4.13). Since ink spreads out isotropically, $\langle R^2\rangle$ is nearly equal to the area covered by ink (see Fig. 4.2). In other words, (4.13) implies that the rate of increase of cross-sectional area as roughly indicated by the spread of ink is proportional to the 4/3 power of its diameter. We may conclude that the fractal dimension of a *sumi-nagashi* is 4/3. The value $4/3 = 1.333$ nearly agrees with 1.35 and 1.3, the experimental values for clouds and *sumi-nagashi* respectively. Corrections to Richardson's law by the fractal structure of turbulence are discussed in the literature [72], but we shall not go into details here.

4.2 Random walk on fractals

4.2.1 *Spectral dimension*

Does any statistical property of random walks on a fractal depend on its fractal dimension? For example, from an engineering standpoint the

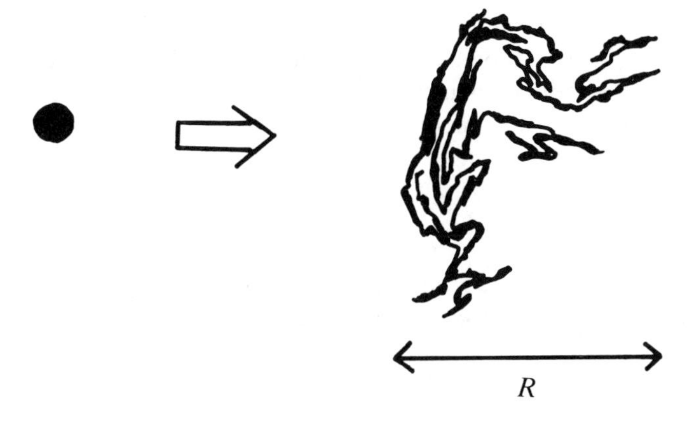

Figure 4.2 Turbulent diffusion.

motion of an electron in an amorphous metal forming fractal percolation cluster is an important problem, so we should seek to solve it by some means. Many scientists have tackled this problem. As a result, it has been determined that random walks on a fractal have several new properties which have never been found in Euclidean space. These new properties are characterised by a new quantity, the so-called *spectral dimension*.

Here we restrict ourselves to consider only discrete random walks on a lattice which move at each discrete time step Δt. We cannot set up the problem with continuous walks because the ordinary diffusion equation does not apply. Diffusion equations are partial differential equations and are the basis for continuous random walks. We cannot construct such equations on any fractal structure because of the non-differentiability of fractals.

On the other hand, in discrete space and time, random walks are well-defined and can be analysed not only numerically but als theoretically by the renormalisation group method, for example. We expect that the results obtained will be independent of the specific discretisation chosen, in the large-scale limit. Hence in the following discussion we shall consider the asymptotic behaviour of random walks on fractals for large time steps [73].

The expectation of the mean square distance from the origin after N time steps $\langle R^2 \rangle$, is expressed as

$$\langle R^2 \rangle \propto N^{\tilde{D}/D} \tag{4.14}$$

where D is the fractal dimension of the fractal structure and $\tilde{D}$ is the spectral dimension for the random walk. In Euclidean space with integer D, we have $\tilde{D} = D$, and (4.14) gives a well-known relation for Brownian random walks. For fractal structures with non-integral D, $\tilde{D}$ is not equal to D, leading to the anomalous behaviour of random walks on fractals.

The total number of distinct sites visited by the walker within N steps, s_N, depends only on the spectral dimension:

$$\langle S_N \rangle \propto \begin{cases} N^{\tilde{D}/2}, & \tilde{D} < 2, \\ N & \tilde{D} \geq 2, \end{cases} \tag{4.15}$$

And $P_0(N)$, which denotes the probability of returning to the origin at time step N, is also dependent on $\tilde{D}$ only:

$$P_0(N) \propto N^{-\tilde{D}/2}. \tag{4.16}$$

The time step $T(\xi)$ at which a particle starting from the origin attains the distance ξ for the first time is given as follows:

$$T(\xi) \propto \xi^{2D/\tilde{D}}. \tag{4.17}$$

This can be regarded as an inversion of (4.14).

These relations can be regarded as natural extensions from Euclidean spaces to fractal spaces. (Equations (4.14) to (4.17) are valid in Euclidean space, where $\tilde{D} = D = d$.)

We have given the above results without clarifying the definition of $\tilde{D}$; it cannot be specified by the fractal dimension D alone, since it is a quantity containing information about temporal behaviour. We may regard any one of the above relations (4.14) to (4.17) as a definition of $\tilde{D}$, since each of them prove to give an identical value. For the purpose of determining $\tilde{D}$ by computer simulation, (4.17) is most effective.

For some fractals we can analytically obtain the spectral dimension by analysing the transition probability by the renormalisation group technique. For example, the spectral dimension of the Sierpinski gasket in d-dimensional space is given as

$$\tilde{D} = \frac{2 \log (d+1)}{\log (d+3)}, \quad \left(D = \frac{\log (d+1)}{\log 2} \right). \tag{4.18}$$

In the case of percolation clusters, $\tilde{D}$ is characterised by critical exponents as

$$\tilde{D} = \frac{2(d\nu - \beta)}{\mu - \beta + 2\nu} \tag{4.19}$$

where ν and β are the critical exponents introduced in Section 2.4, and μ is the exponent for electric conductivity σ:

$$\sigma \propto (\rho - \rho_c)^{\mu}. \tag{4.20}$$

The quantity $\mu-(d+2)\cdot\nu$ is known to be nearly equal to 1. Hence we get the result:

$$\widetilde{D} \doteqdot \frac{4}{3}. \tag{4.21}$$

This suggests an interesting universality that random walk on percolation clusters depend only very slightly on the embedded space.

The reader may agree that the spectral dimension is an important quantity for random walks on fractals, but may wonder why $\widetilde{D}$ is called the spectral dimension. The name comes from another property of $\widetilde{D}$, as follows.

Let us consider an elastic material with fractal structure such as a Sierpinski gasket or percolation cluster. If it is slightly bent, it will oscillate around its equilibrium shape. The spectral density $\rho(\omega)$ of the oscillation may be expressed as a function of frequency, ω, as

$$\rho(\omega) \propto \omega^{\widetilde{D}-1} \tag{4.22}$$

Since the quantity $\widetilde{D}$ governs the spectrum of oscillations of the fractal structure, it is named the spectral dimension. It is not trivial that properties concerning diffusion, (4.14)–(4.17), and the property about oscillation, (4.22), are characterised by only one parameter. In fact, some doubts have recently been raised – for details see [74].

Another aspect of the spectral dimension can be found in a problem about self-avoiding random walks on a fractal: is the formula $D' = (d+2)/3$ applicable when the space dimension d is not an integer? The answer is no, and we need a new formula for this problem, given as follows [75]:

$$D' = \frac{D}{\widetilde{D}} \cdot \frac{\widetilde{D}+2}{3}, \quad (\widetilde{D} \leq 4). \tag{4.23}$$

Mandelbrot [1] has shown that the spectral dimension can be expressed by more elementary fractal dimensions (see Section 6.2).

4.2.2 *Long time tails*

In Section 2.4 we saw that there are many phenomena in nature which exhibit long time correlations obeying power laws. Here, we introduce some examples with long time tails caused by random walk properties.

A system where many particles scatter, interacting electrically on each other, may be regarded as a model of a gas. In such a system, each particle is scattered successively, so it walks almost randomly. Although velocity autocorrelation of such a particle was believed to decay exponentially, Alder and Wainwright discovered a long time tail by computer simulation [76]. This discovery led to a number of studies on long time tails and it is now clear that the correlation function can be expressed as

$$\langle v(0)\, v(t) \rangle \propto t^{-d/2},\; d \geq 2, \tag{4.24}$$

where d denotes the dimension of the space. This relation comes from the viscosity effect that an external impact is relaxed by viscosity following the same power law as (4.24).

The existence of long time tails indicates that particles in a gas do not forget their history. Their walks should be distinguished from Markovian Brownian motion. We often use the Fokker–Planck equation or the following generalised version of it in order to describe motions of diffusing particles:

$$\frac{\partial}{\partial t} S(x, t) = \sum_{j=1}^{\infty} a_j \frac{\partial^j}{\partial x^j} S(x, t) \tag{4.25}$$

where S(x, t) denotes the probability density that a particle can be found on x at time t. However, this equation is not valid for random walkers with long time tails, because diffusion coefficient a_2 or higher order coefficients, a_3, a_4, ... must be divergent for such a random process. Therefore the problem of long time tails plays a crucial role in the statistical properties of transport processes.

Long time tails of random walks are also found in a much simpler system called the *Lorentz gas model* where a classical particle is scattered elastically off randomly located fixed scatterers. This model was originally introduced as a model of electron motion in metal, but readers may regard it as a model of pinball.

In the case where the scatterers line up regularly to form a lattice, it is proved that the velocity autocorrelation of the particle decreases faster than any power of time, namely, there is no long tail. On the contrary, if the scatterers are distributed at random, it is known that there exists a long time tail of the form [77]:

$$\langle v(0)\, v(t) \rangle \propto -\, t^{-\left(\frac{d}{2}+1\right)} \tag{4.26}$$

in d-dimensional space. Compared with the long tail of (4.24), we find two differences: the power exponent and the sign of the tail. In the Lorentz gas model, the correlation decreases so rapidly that it transverses zero in a short time regime, then it relaxes, gradually obeying a power law.

An interesting point about these long time tails may be the exponent's dependence on spatial dimension, d. A question naturally arises from the standpoint of fractals: Can we extend this value of dimension to non-integers? An answer to this question will now be given for a special case [78].

Let us consider a stochastic version of the Lorentz gas model in one-dimensional space. This model consists of a particle moving with velocity $\pm c$ along the x-axis, and at every lattice point $x = i\Delta$ (i an integer) there is a point-like scatterer. The particle which is located in the interval $((i-1)\Delta, i\Delta]$ at $t = n\,\Delta/c$ with velocity $+c$ will be found either in the adjacent interval $(i\Delta, (i+1)\Delta]$ with velocity $+c$ or in $((i-1)\Delta, i\Delta]$ with the opposite velocity $-c$ at $t = (n+1)\Delta/c$. The corresponding probabilities of occurrence are given by $1 - e^{-a_i}$ and e^{-a_i}, respectively. Thus the particle is elastically reflected with probability $1 - e^{-a_i}$ by the scatterer at $x = i\Delta$ (see Fig. 4.3). The positive number, a_i, designates the strength of the scatterer at the point $x = i\Delta$. For this system, the following set of difference equations holds strictly:

$$\begin{aligned} S_i^+(n+1) &= e^{-a_i} \cdot S_{i-1}^+(n) + (1-e^{-a_i}) \cdot S_i^-(n), \\ S_i^-(n+1) &= (1 - e^{-a_{i+1}}) \cdot S_i^+(n) + e^{-a_{i+1}} \cdot S_{i+1}^-(n), \end{aligned} \tag{4.27}$$

where $S_i^+(n)$ and $S_i^-(n)$ denote the probability densities that the particle is located in the interval $(i\Delta, (i+1)\Delta]$ at $t = n\Delta/c$ with the velocities $+c$ and $-c$, respectively.

The velocity autocorrelation is expressed as

$$\left\langle v\left(\frac{n\Delta}{c}\right) \cdot v(0)\right\rangle = c^2 \frac{\Sigma\,(S_i^+(n)_i - S_i^-(n))}{\Sigma\,(S_i^+(0)_i - S_i^-(0))}$$

Here the initial condition for (4.27) is $S_i^+(0) = -S_i^-(0) = \text{const}$. By this initial condition, an average is taken over the initial position and direction of the particle. A negative value of the probability S may look unphysical, but this is just a mathematical trick and there is no problem.

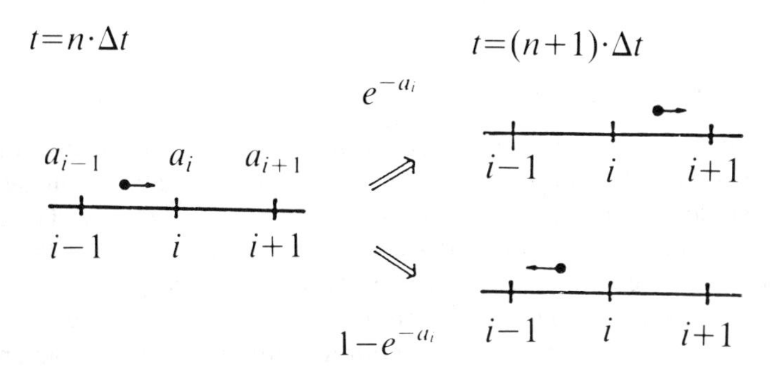

Figure 4.3 Possible motions of particle in one-dimensional Lorentz gas.

In the case that the scatterer distribution is fractal with dimension D, $\{a_i\}$ satisfies the following relation:

$$\overline{(a_j - a)(a_i - a)} \propto \begin{cases} |i-j|^{D-1}, & 0 < D < 1, \\ \delta_{ij}, & D = 0, \end{cases} \tag{4.29}$$

where the bar designates the ensemble average and $a = \bar{a}_i$ (independent of i). Satisfying this relation, each a_i is determined randomly.

An example of the results is shown in Figure 4.4 for the cases $D = 0$ and $D = 0.6$, We see the long time tails in the region $t > 50$. The exponents of the power of the long time tails are -1.50 and -1.21 for $D = 0$ and $D = 0.6$, respectively. Although the exponent for $D = 0$ agrees well with the value expected from (4.26), $-3/2$, the exponent for $D = 0.6$ obviously differs from $-3/2$. Examining the exponents for various values of D, it is confirmed that (4.26) can be extended as follows in the case of where the spatial dimension is 1 and the impurity has fractal dimension D.

$$\langle v(t)\, v(0)\rangle \propto -t^{-\left\{\frac{1-D}{2}+1\right\}}; \quad 0 \le D < 1. \tag{4.30}$$

This result has been confirmed both numerically and theoretically. (Note that white random distribution of impurity corresponds to D=0, hence (4.26) applies in such case.)

Thus, the long time tail of the stochastic Lorentz gas model in 1-dimensional space has been extended to the fractal case. In d-dimensional space, the following long tail is expected to appear:

$$\langle v(t)\, v(0)\rangle \propto -t^{-\left\{\frac{d-D}{2}+1\right\}}, \tag{4.31}$$

where D is the fractal dimension of scatterers. (Note: the quantity $d-D$ is called the co-dimension.)

4.3 The devil's staircase

Spin systems with short-range interactions have been intensively studied; however, relatively little knowledge has been obtained about spin systems that have long-range interactions such as Coulomb forces. Here, we introduce a long-range spin model where the relation between the external field and susceptibility produces a Devil's staircase [79].

Let us consider a spin system on a 1-dimensional lattice where each spin s_i takes the value ± 1. The Hamiltonian of this system is given as

$$\mathcal{H} = \sum_{i=1}^{N} H \cdot S_i + \frac{1}{2} \sum_{i,j=1}^{N} J(i-j)(S_i + 1)(S_j + 1), \tag{4.32}$$

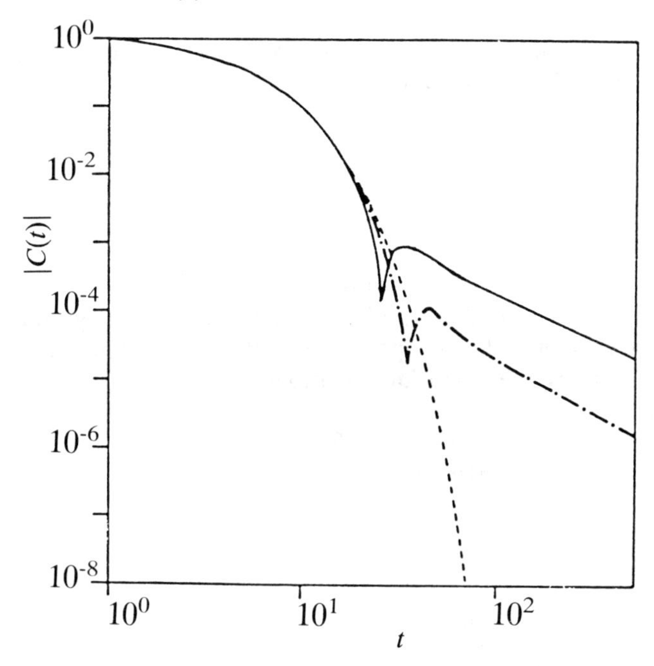

Figure 4.4 The absolute value of the velocity autocorrelation v. time: $D = 0$ (dash-dotted line); $D = 0.6$ (solid line).

where H denotes the external field, and the interaction kernel $J(i - j)$ satisfies the following relations:

(1) $J(i) \to 0$ as $i \to \infty$;
(2) for all i, $J(i + 1) - 2J(i) + J(i - 1) \geq 0$;
(3) $J(i) > 0$ (antiferromagnetic).

The second condition says that the interaction is a concave function of distance. Note that in this system we are considering only interactions between up-spins ($s = +1$) since down-spins do not contribute to the second term in (4.32). Hence, we may regard this system as a set of identically charged particles having repulsive interactions such as Coulomb interactions.

When the fraction q of up-spins is given, the ground state and the lowest energy state are obtained as follows. In the case $q = 1/n$, where n is a natural number, between any two successive up-spins there are $n - 1$ down-spins. In other words, up-spins are located at every nth spin. In a more general situation $q = m/n$ where m and n are irreducible integers, the ground state has a periodic spin configuration with period n containing m up-spins in each period. In this case, distances between two neighbouring up-spins are either $[n/m]$ or $[n/m] + 1$ (where the bracket

$[x]$ gives the largest integer that does not exceed x). For example, when $q = \frac{1}{3}, \frac{2}{5}, \frac{3}{7}$, the spin configurations become

$$\begin{aligned} &\tfrac{1}{3}; \ \ldots \ +--+--+-- \ \ldots \\ &\tfrac{2}{5}; \ \ldots \ +--+-+--+- \ \ldots \\ &\tfrac{3}{7}; \ \ldots \ +-+-+--+-+- \ \ldots \end{aligned} \tag{4.34}$$

where + and − denote up-spin and down-spin, respectively.

For a given rational number q, it has been proved that the ground state can be obtained by the following procedure [80]. First, define the integers k, n_0, n_1, ..., n_k by the following equations:

$$\begin{aligned} 1/q &= n_0 + r_0 \\ |1/r_0| &= n_1 + r_1 \\ &\vdots \\ |1/r_{k-2}| &= n_{k-1} + r_{k-1} \\ |1/r_{k-1}| &= n_k \end{aligned} \tag{4.35}$$

for all j, $-\frac{1}{2} < r_j \le \frac{1}{2}$ (the sequence must terminate for rational q). Second, define the sequences $X_1, X_2, \cdots, X_k$ and $Y_1, Y_2, \cdots, Y_k$ by

$$\begin{aligned} X_1 &= n_0 \\ Y_1 &= n_0 + \alpha_0 \\ X_{i+1} &= (X_i)^{n_i - 1} \cdot Y_i \\ Y_{i+1} &= (X_i)^{n_i + \alpha_i - 1} \cdot Y_i \end{aligned} \tag{4.36}$$

where $\alpha_i = r_i/|r_i| = \pm 1$.

Then X_{k+1} gives the spin configuration of the lowest energy. For example, consider the case of $q = 11/47$. From (4.35) we have $n_0 = 4$, $n_1 = 4$, $n_2 = 3$ and $\alpha_0 = 1$, $\alpha_1 = -1$. And from (4.36), we get $X_1 = 4$, $Y_1 = 5$, $X_2 = 4^3 \cdot 5$, $Y_2 = 4^2 \cdot 5$ and $X_3 = (4^3 \cdot 5)^2 \cdot 4^2 \cdot 5$. Here the superscripts indicate the numbers of repetitions and X_3 can be also written as

$$X_3 = 4 \cdot 4 \cdot 4 \cdot 5 \cdot 4 \cdot 4 \cdot 4 \cdot 5 \cdot 4 \cdot 4 \cdot 5. \tag{4.37}$$

In this expression, the number 4 (or 5) shows that the 4th (or 5th) spin counting from an up-spin is up. The sum of these numbers 4 and 5 is 47, which means the period of the spin configuration is 47. As X_2 is composed of 11 figures we find that there are 11 up-spins in a period. The above-mentioned example (4.34) can be expressed in this form as

$$\tfrac{1}{3} \to 3, \quad \tfrac{2}{5} \to 3 \cdot 2, \quad \tfrac{3}{7} \to 2^2 \cdot 3. \tag{4.38}$$

Now we know the spin configuration in the ground state, let us investigate how the ratio of up-spin, q, will change when the external field is changed. For this purpose, we first consider the change of total energy when one up-spin in a ground state with $q = m/n$ is flipped to a down-spin. In this case, the change of energy ΔU is given as follows.

$$\begin{aligned} \Delta U = 2H &+ 4(r_1 + 1)\, J(r_1) - 4r_1\, J(r_1 + 1) \\ &+ 4(r_2 + 1)\, J(r_2) - 4r_2\, J(r_2 + 1) + \ldots \\ &+ 4nJ\,(n - 1) - 4(n - 1)\, J(n) + \ldots \\ &+ 4 \cdot 2nJ(2n - 1) - 4(2n - 1)\, J(2n) + \ldots, \end{aligned} \tag{4.39}$$

where r_i denotes the interval between the flipped spin and the ith up-spin, and is given by a natural number which satisfies the following relation:

$$r_i \leq \frac{n}{m}\, i < r_i + 1. \tag{4.40}$$

In (4.39), the relations $r_m = n$, $r_{2m} = 2n$, have already been substituted. In the same way, the change of total energy when one down-spin is flipped to up with the same q and external field H' is given by

$$\begin{aligned} \Delta U' = -\, 2H' &- 4(r_1 + 1)\, J(r_1) + 4r_1\, J(r_1 + 1) \\ &- 4(r_2 + 1)\, J(r_2) + 4r_2\, J(r_2 + 1) - \ldots \\ &- 4(n + 1)\, J(n) + 4nJ(n + 1) - \ldots \\ &- 4(2n + 1)\, J(2n) + 4 \cdot 2nJ(2n + 1) - \ldots. \end{aligned} \tag{4.41}$$

In a stable state where total energy takes a minimal value, the total energy will change only slightly if we change the direction of a single spin. Therefore ΔU and $\Delta U'$ should be nearly equal to zero for the stable configuration. From these conditions, the change of the external field $\Delta H = H - H'$ is given as

$$\begin{aligned} \tfrac{1}{2}\Delta H\left(q = \frac{m}{n}\right) = n &\{J(n + 1) + J(n - 1) - 2J(n)\} \\ &+ 2n\,\{J(2n + 1) + J(2n - 1) - 2J(2n)\} + \ldots \\ &+ in\,\{J(in + 1) + J(in - 1) - 2J(in)\} + \ldots. \end{aligned} \tag{4.42}$$

This equation determines the change of H at $q = m/n$, hence the relation between q and H can be calculated immediately. It is remarkable here that the right-hand side of this equation is independent of m. Since $J(i)$ is a concave function, the right-hand side of (4.42) is positive and finite for any n and m. Thus, if we change q continuously, there appears a finite positive gap at each rational q. This relation between q and H makes a devil's staircase as shown in Figure 4.5, where $J(i) = i^{-2}$. If we enlarge any part of this staircase, we find a similar staircase and, of course, this is a fractal.

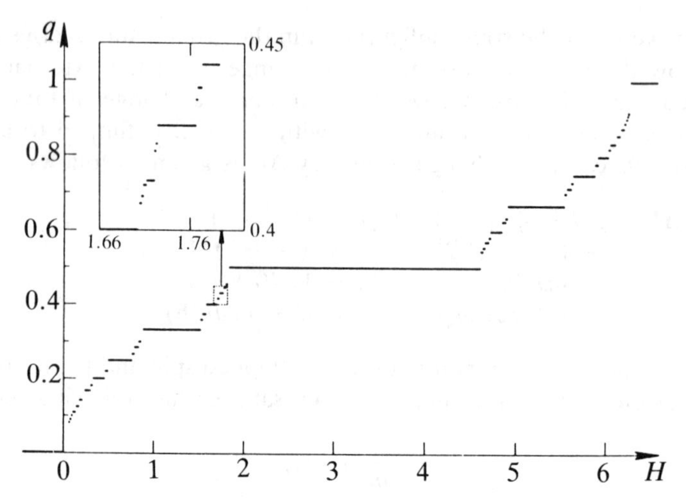

Figure 4.5 Magnetisation (number of up-spins) v. external field H.

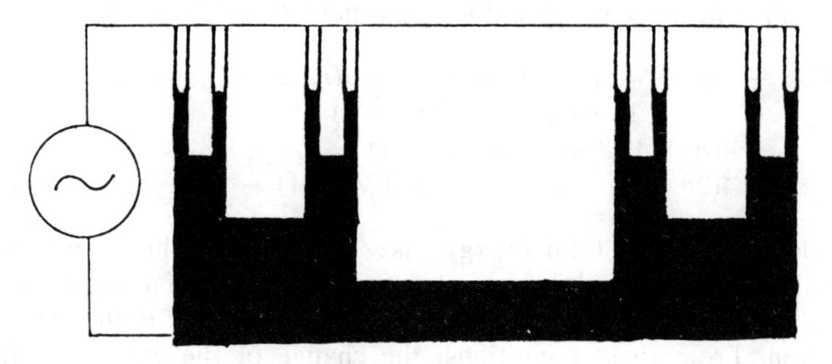

Figure 4.6 An example of a fractal surface.

4.4 AC Response of fractal surface

Electrical properties of fractal surfaces pose very interesting problems. Let us consider a model of a fractal surface based on the Cantor set as shown in Figure 4.6 [81]. As seen from the figure, each groove has two branches, and each branch is similar to the whole groove when magnified by a factor a, $a > 2$. The dimension of this Cantor set is $\log 2/\log(a) < 1$. The surface has the dimension $2 + \log 2/\log(a)$.

Branching of an electric current repeats when the current flows from bottom to top. The resistance R of each branch increases by the ratio a at every stage of branching because it is proportional to the cross-sectional area. The interfaces in the groove facing each other make a condenser of capacitance C, which is assumed to be the same at every stage. Using self-similarity, we can then show that the impedance of this surface $Z(\omega)$ satisfies the following equation:

$$Z\left(\frac{\omega}{a}\right) = R + \frac{aZ(\omega)}{-i\omega cZ(\omega) + 2} \tag{4.43}$$

In the low-frequency limit this reduces to

$$Z\left(\frac{\omega}{a}\right) \doteqdot \frac{a}{2} Z(\omega). \tag{4.44}$$

The solution of this equation becomes

$$Z(\omega) \propto \omega^{-\alpha}, \tag{4.45}$$

where $\alpha = 1 - \log 2/\log(a)$. If we write this exponent generally as

$$\alpha = 3 - D, \tag{4.46}$$

then D is the fractal dimension of the surface. Thus we have an inverse power-law frequency dependence for a fractal surface. This result may have great importance in technological problems.

4.5 Model of a river and fractal distribution

Here we introduce a very simple model of a river which shows a power law distribution [82]. By analysing and generalising this model the origin of the power law will be clarified.

First, let us assume that rain is falling at a constant rate and uniformly on a slope. If the surface of the slope is not flat but has random irregularities, then rain will run down the slope with random fluctuations. When two rain drops collide with each other, they will join and make one drop which runs randomly just like before the collision. The trajectories of random walks are regarded as tributary streams, and collision points are regarded as confluences.

In Figure 4.7, an example of a river pattern is shown. We can find many dendritic structures densely distributed on the lattice. If we consider the drifting direction as the time axis, this system becomes an aggregating system in 1-dimensional space with uniform injection of smallest-size particles.

One of the most important quantities in this system is the distribution of flow rates far down-stream. By numerical simulation, it is confirmed that the size distribution converges to the following power law, independent of initial conditions:

$$P(S) \propto S^{-\alpha}, \quad \alpha = \frac{1}{3}. \tag{4.47}$$

Figure 4.7 An example of a river pattern.

This distribution corresponds to the distribution of particle size in the aggregation system.

Equation (4.47) can be explained intuitively as follows. The rate of flow of a river is proportional to the area of its drainage basin. Since the drainage basin is surrounded by ridges as shown in Figure 4.8 and the ridges are graphs of Brownian motions, we have to estimate the distribution of the area surrounded by two Brownian motions. Roughly speaking, the area is proportional to the product of the length and the width of the river. The distribution of the length (denoted by L) is identical to the distribution of recurrence time (see Section 5.2); hence we have

$$P(L) \propto L^{-1/2} \tag{4.48}$$

It is plausible to assume that the width is nearly proportional to $\sqrt{L}$ and therefore the area S is proportional to $L^{3/2}$. Then substituting $L \propto S^{2/3}$ into (4.48) we obtain (4.47).

The essence of the above power law is random aggregation and constant injection (rainfall). In fact if we introduce divisions of rivers or if we stop the rain, the distribution changes form and decays faster than any power at large S. However, the power law is stable for other perturbations.

The situation becomes clearer by considering a mean field version of this model. Assume that there are N sites on a lattice and at each site there is one particle of size s_j. Each particle jumps to a randomly chosen site. If two or more particles come together at a site, they are united to form a new particle with conserved mass. When particles with unit mass are added at all the sites they are absorbed by any particles already there.

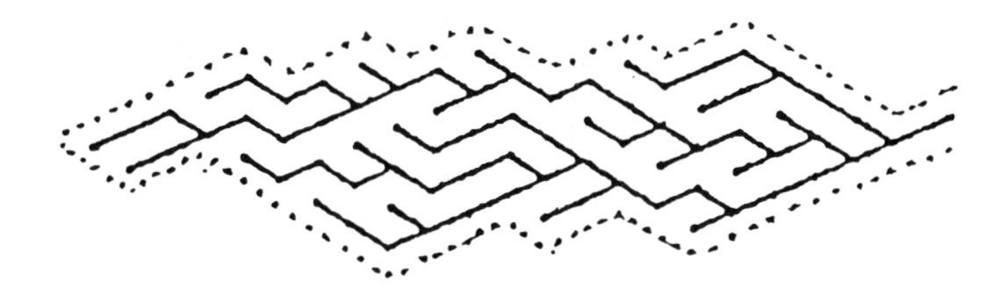

Figure 4.8 Rivers and ridges.

This procedure can be expressed by the following equation:

$$s_i\,(n+1) = \sum_j W_{ij}\,(n)\, s_j\,(n) + 1 \tag{4.49}$$

where $W_{ij}(n)$ denotes the realisation of the particle at the jth site jumping to the ith site at time step n and is given as

$$W_{ij}(n) = \begin{cases} 1, & \text{probability } 1/N, \\ 0, & \text{probability } 1 - 1/N \end{cases} \tag{4.50}$$

By repeating this procedure many times we have the following distribution for s:

$$P(S) \propto s^{-\alpha}, \; \alpha = 1/2. \tag{4.51}$$

Time evolution of the distribution of particle sizes is governed by the following equation for the probability density $p(s,n)$,

$$\left(P(s) = \int_s^\infty p(s')\, \mathrm{d}s' \right)$$

$$p(s+1, n+1) = \sum_{r=1}^{N} a_r \sum_{s_1+s_2+\,.\,.\,+s_r=s} \prod_{i=1}^{r} p(s_i, n), \tag{4.52}$$

with boundary condition

$$p(1, n) = a_0, \quad \text{for all } n \geq 0, \tag{4.53}$$

where the coefficient a_r denotes the probability that r particles come together at a site and is given by

$$ar = {}_N\mathrm{C}_r \left(\frac{1}{N}\right)^r \cdot \left(1 - \frac{1}{N}\right)^{N-r}. \tag{4.54}$$

The asymptotic behaviour of this equation is obtained by introducing the characteristic function

$$Z\,(\rho, n) = \sum_{s=1}^{\infty} \mathrm{e}^{-\rho s}\, p(s, n). \tag{4.55}$$

The equation for Z at $n=\infty$ becomes the following algebraic equation:

$$Z(\rho) = e^{-\rho} \left\{ \sum_{r=0} a_r + \sum_{r=0} r \cdot a_r (Z - 1) \right.$$
$$\left. + \sum_{r=2}^{N} \frac{r(r-1)}{2} a_r \cdot (Z - 1)^2 + \dots \right\}. \qquad (4.56)$$

In the vicinity of $\rho = 0$, we can neglect higher order terms in $Z - 1$ because $Z(0) = 1$ from the normalisation of probability. If the coefficient of the second term in the right-hand side, $\sum_{r=0}^{\infty} r\ a_r$ is not 1, we have $Z(\rho) - 1 \propto \rho$ which indicates exponential decay in the distribution function. However, in our problem, the coefficient is identically 1, because the number of particles is kept constant owing to the uniform injection. Therefore the second term cancels the left-hand side and the third term becomes dominant. Then we have $Z(\rho) - 1 \propto \rho^{1/2}$ which corresponds to (4.48).

From the above results we can expect the following general statement: In an aggregating system with constant injection, the effects of injection and aggregation are balanced and a non-trivial quasi-stationary size distribution is realised in the large time limit. The distribution has a fractal property and obeys a power law [82].

This statement gives an intuitive answer to the question as to why there are so many fractals in nature. Most physical systems in nature are dissipative. Thus, without injection they cannot keep stationary and tend to be in a trivial state. With constant injection, dissipation is balanced with the injection, and the system becomes a quasi-stationary state where fractal properties are likely to appear.

Note

1 The Holtzmark distribution has been observed in high-density ion beams from conventional plasma ion sources. The long tails in the transverse velocity distribution impose a limitation in the focusing of the beam [116].

5 Mathematical methods for fractals

This chapter is devoted to some mathematical methods which are useful for the treatment of fractals. Preliminary knowledge of advanced mathematics is not required here. As seen in the previous chapters, we need only elementary analysis. The renormalisation group technique in Section 5.1 and the dimensional analysis in Section 5.3 are too simple to be called 'mathematical' – rather it would be appropriate to regard the present topics as useful tools in physics.

5.1 The renormalisation group

One might have the impression that the theory of the renormalisation group [83] is very difficult, because Wilson was awarded the Nobel prize for physics in 1982 in honour of his contribution to renormalisation group theory. But in fact this is not so: at least its essence can be understood with high school mathematics. In particular, the real space renormalisation group [84], which is presented later with examples, is easy to understand and is deeply related to fractals.

The purpose of the renormalisation group is to treat quantitatively the change of a physical quantity when the degree of coarse-graining in observation is changed. For example, let p be a certain physical quantity measured at a certain scale of coarse-graining, and let the same quantity measured with the scale of coarse-graining doubled be p'. This p' may be related to the original value p by an appropriate 'coarse-graining' transformation f_2 such as:

$$p' = f_2(p), \tag{5.1}$$

where the subscript 2 of f denotes the ratio of coarse-graining. If the scale of coarse-graining is again doubled, then

$$p'' = f_2(p') = f_2 \cdot f_2(p) = f_4(p). \tag{5.2}$$

Generalising these relations, we see that the transformation f has the following properties:

$$f_a \cdot f_b = f_{ab}, \tag{5.3}$$

$$f_1 = 1, \tag{5.4}$$

where 1 denotes an identity transformation. For any given state, we can easily define its coarse-grained state. However, the inverse process, to find the original state which is coarse-grained to the given state, cannot be determined uniquely. In fact, the transformation f does not generally have an inverse f^{-1}. The set of transformations having these properties is mathematically called a semi-group. And the transformation of coarse-graining is called renormalisation in physics, so this transformation f should rigorously be called 'renormalisation semi-group'. But the name '*renormalisation group*' is now conventional.

This renormalisation group is closely related to fractals as can be seen from the definition. Since a fractal is an object that remains unchanged through coarse-graining, it can be said that a fractal is an object that is invariant under the transformation of the renormalisation group. Historically, the notions of fractal and renormalisation group were born independently at almost the same time. Both were intended to analyse what is invariant under the change of scale of observation: fractal for geometrical objects and renormalisation group for physical quantities. Recently, fractals have come to include physical quantities and the real-space renormalisation group has come to treat geometrical objects, so the difference between them is becoming less.

The renormalisation group is a powerful tool for the analysis of critical phenomena in phase transitions. For example, let us consider a state of water near the critical point between liquid phase and gas phase, as described in Section 2.4. If the considered state is in the liquid phase slightly below the critical point, then this state is microscopically a random mixture of liquid and gas. At each step of coarse-graining, those regions where the liquid is dominant are observed as liquid, and the ratio of liquid will increase with the degree of coarse-graining. The limit state after repeating this process infinitely many times will be observed as complete liquid. Inversely, if there are more gas regions than liquid regions in the microscopic state, we will get a complete gas state in the macroscopic limit.

Let us analyse the critical phenomena of percolation by using the renormalisation group technique. Consider a 2-dimensional square lattice with randomly distributed metal atoms on lattice points. As a physical quantity, p, we take the probability that a lattice point is occupied by a metal atom.

Now we consider the 2×2 lattice points as a single virtual lattice point by performing coarse-graining. We call this new lattice a super-lattice and the 2 × 2 lattice points which have been lumped together a block. If there are metal atoms at all four points in the block, we regard the corresponding super-site (the site on the super-lattice) as occupied by a metal atom. If three points in a block are occupied by metal atoms, we may also regard this as a metal atom on the super-site, since the block conducts to both directions, vertical and horizontal. But when less than three lattice points in a block are occupied by metal atoms, this block only conducts at most in one direction. So in this case, we should regard the super-site as unoccupied (see Fig. 5.1). Thus, the probability that there exists a metal atom on a super-site can be given by

$$p' = f_2(p) = p^4 + 4p^3(1 - p). \tag{5.5}$$

The first and second terms respectively correspond to the cases that four and three points in a block are occupied by metal atoms. Now we have determined the transformation f_2, we can analyse the phase transition by studying the properties of f_2.

As mentioned before, the critical point p_c is the point which is in-

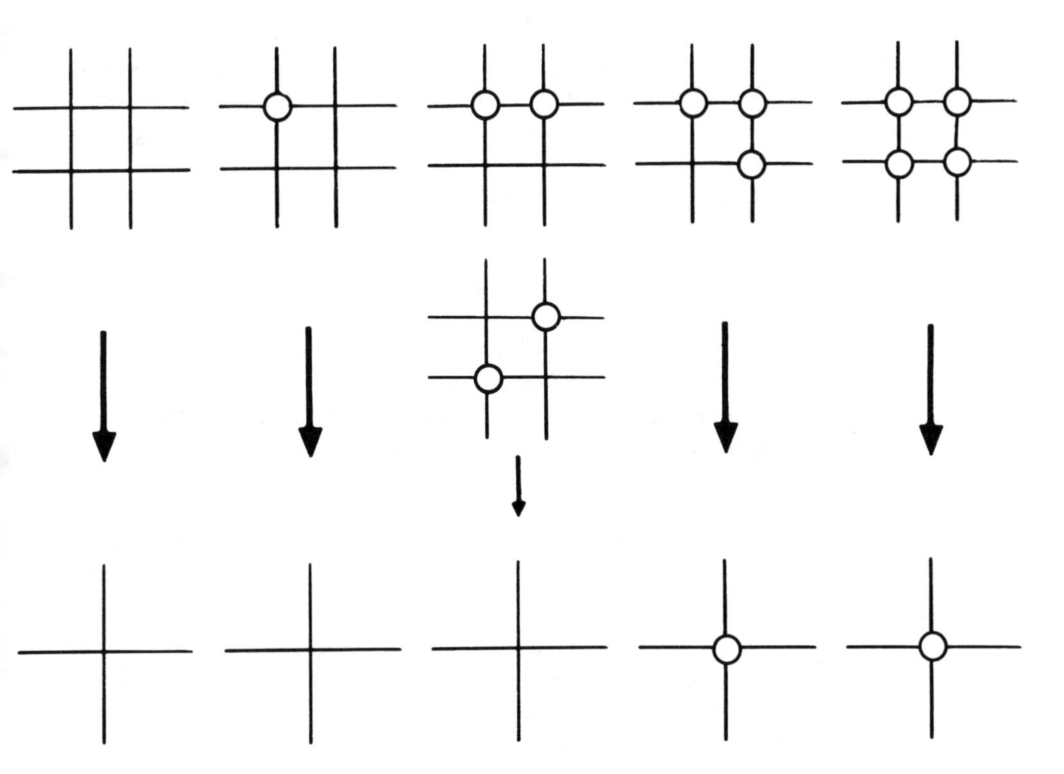

Figure 5.1 Renormalisation from a block to a super-site.

variant under the transformation f_2, that is, a fixed point. Denoting the fixed point by p^*, then from (5.5), we have

$$p^* = p^{*4} + 4p^{*3}(1 - p^*), \tag{5.6}$$

and we get

$$p^* = 0,\ 1,\ \frac{1 \pm \sqrt{13}}{6}. \tag{5.7}$$

and $\dfrac{1 \pm \sqrt{13}}{6}$ evaluates as -0.434 and 0.768.

Since p is a probability, the negative value $p^* = -0.434$ must be discarded. The p^* points $= 0$ and 1 are trivial fixed points which correspond to the cases that all lattice points are occupied or unoccupied by metal atoms, respectively. Thus, the critical value at which phase transition occurs should be $p_c = 0.768$. This value is a little greater than $p_c = 0.59$ obtained from computer simulation, but approximately agrees with the experimental value $p_c = 0.752$.

For $p < p_c$, the following inequalities hold:

$$p_c > p > f_2(p) > f_{2^2}(p) \ldots > f_{2^n}(p), \tag{5.8}$$

that is, p is transformed to be smaller at every step of renormalisation, and after repeating an infinite number of times, $f_\infty(p) = 0$. This means that an occupied site cannot be found in the infinitely coarse-grained state.

For $p > p_c$ the converse is the case and $f_\infty(p) = 1$, that is, in the limit of coarse-graining, the lattice becomes totally metallic. In this way, points near p_c become more distant from p_c at each step of renormalisation, so the critical point p_c is an unstable fixed point of the transformation f_2 (Fig. 5.2).

Now we have the critical point p_c, let us next calculate the fractal dimension of the percolation clusters at p_c. When a super-site is metallic, three or four points in the block are metal. The expectation value of the number of metallic lattice points in this block is

$$\begin{aligned} N_c &= \{4 \cdot p_c^4 + 3 \cdot 4 \cdot p_c^3(1 - p_c)\}/p_c \\ &\fallingdotseq 3.45. \end{aligned} \tag{5.9}$$

Here, we have divided by p_c because N_c is the expectation value under the condition that this super-site is metal. The lattice spacing is half of the super-lattice's spacing. That is, one metallic atom in a super-lattice can be seen as N_c metallic atoms on the average when the scale of observation is halved. Generalising this relation, the number of metallic atoms seen when the scale of observation becomes $1/b$ can be expressed as

$$N_c(b) = b^{-D}. \tag{5.10}$$

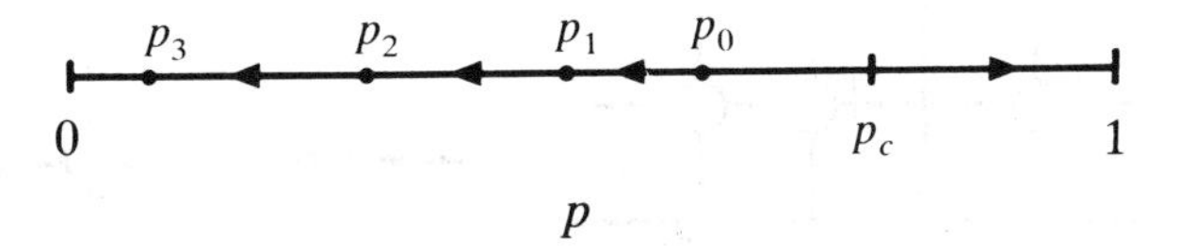

Figure 5.2 The change of p by renormalisation.

In this case $b = 2$, and we obtain the fractal dimension as

$$D = \frac{\log N_c}{\log 2} \doteqdot 1.79. \tag{5.11}$$

This D gives the fractal dimension of a percolation cluster. The value $D = 1.79$ is in good agreement with that obtained by computer simulation and close to the experimental value 1.9. This shows the excellence of the renormalisation group method.

We have calculated the fractal cluster dimension just at the critical point. Other characteristics near the critical point, especially the critical exponents, are readily found by using the renormalisation group. As an example, let us calculate the critical exponent for the correlation length.

When p is smaller than the critical value p_c but sufficiently close to it, the correlation length ξ is finite and can be expressed, as explained in Section 2.4, as follows.

$$\xi = \xi_0 \, |p_c - p|^{-\nu}, \tag{5.12}$$

where the proportional constant ξ_0 has the dimension of length and is of the order of the lattice spacing. Considering the rescaled super-lattice, the following equation also holds since the correlation length ξ is itself invariant under coarse-graining.

$$\xi = \xi'_0 \, |p_c - p'|^{-\nu}, \tag{5.13}$$

where ξ'_0 is given by

$$\xi'_0 = 2\xi_0, \tag{5.14}$$

because the extension rate of the lattice spacing by the rescale is 2. From (5.12)–(5.14), the critical exponent is expressed as

$$\nu = \frac{\log 2}{\log \{(p_c - p')/(p_c - p)\}}. \tag{5.15}$$

If we consider the limit when p and p' approach p_c, then

$$\frac{p_c - p'}{p_c - p} \to \left. \frac{\partial p'}{\partial p} \right|_{p=p_c} \tag{5.16}$$

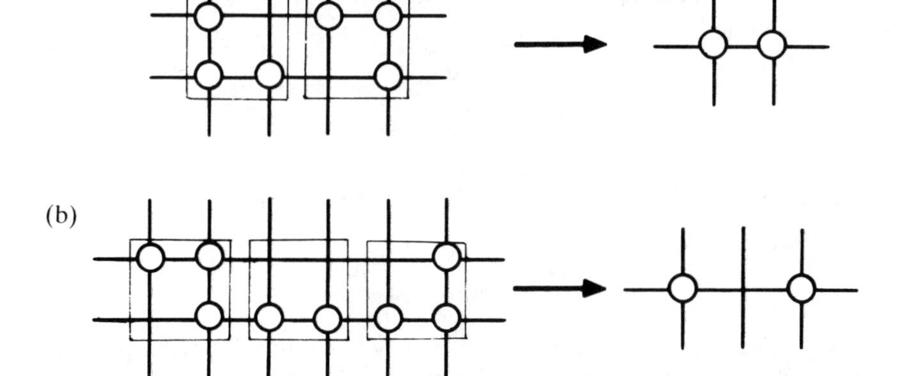

Figure 5.3 Errors in renormalisation: (a) disconnected sites become connected; (b) connected sites become disconnected.

and therefore,

$$\nu = \frac{\log 2}{\log \left. \dfrac{\partial f_2(p)}{\partial p} \right|_{p=p_c}}. \tag{5.17}$$

Using (5.5), we obtain $\nu = 1.40$. This value is nearly equal to $\nu = 1.35$ which has been derived from simulation and experiments.

In this way, with the renormalisation group technique, we can obtain the fractal dimension and critical exponents with relatively simple calculations. However, we should keep it in mind that the renormalisation group technique is an approximation – see Figure 5.3 (a) and (b) for example. In (a), disconnected points in the lattice become connected in the super-lattice, and conversely in (b), connected points in the lattice are regarded as disconnected in the super-lattice. In order to reduce these errors, it is better to take the size of the block as large as possible. If we choose a block with side length b, then b^2 points are in the block and the number of possible states of the block becomes 2^{b^2}. This number increases so rapidly that $b = 4$ is the upper limit for which the transformation f_b can be obtained analytically.

Instead of calculating f_b analytically the Monte Carlo renormalisation method has been developed, which determines f_b statistically. In this method, the renormalised probability p' for a given p is estimated by repeating simulations on, for example, a 100×100 lattice ($b = 100$). In each realisation we distribute metal points randomly with probability p and check whether or not a percolating cluster appears on the lattice.

This procedure is performed for various values of p, and the function f_b is determined approximately. The accuracy of renormalisation becomes higher because larger blocks are renormalised at a time. Once we have determined the renormalisation function, f_b, we can estimate the properties of a very large system which cannot be studied directly even by computer because of memory or CPU time restrictions. This method will become more and more important with the advance of computational physics.

5.2 Stable distribution

We have come across the stable distribution in Sections 2.4 and 4.1. Here we present a more detailed explanation since it is very closely related to fractals.

The notion of stable distribution was invented by P. Lévy in the early 1930s. He studied summation of random variables and found that there are some distributions which reproduce themselves up to a linear change of variable under the operation of convolution. A definition of *stable distribution* [85] is given by Feller as follows:

Let $X, X_1, \ldots, X_n$ be independent random variables with a common distribution R. The distribution R is stable if and only if for $Y_n \equiv X_1 + X_2 + \ldots + X_n$ there exist constants c_n and γ_n such that

$$Y_n \stackrel{d}{=} c_n X + \gamma_n \tag{5.18}$$

where $\stackrel{d}{=}$ indicates that the random variable of both sides have the same distribution.

In general, the sum of random variables with a common distribution becomes a random variable with a distribution of different form. However, for random variables with a stable distribution, an appropriate linear transformation makes the sum of random variables obey the same distribution. In other words, the distribution of the sum of many random variables is similar to that of one variable. This may be regarded as a kind of self-similarity.

The best-known distribution having this property is the Gaussian distribution. (A sum of Gaussian random variables is also Gaussian.) However, a stable distribution does not need to be Gaussian. Rather, the Gaussian distribution is a very special case of stable distribution, as will be seen later.

Using the characteristic function of a distribution

$$\phi(z) \equiv \langle e^{iXz} \rangle, \tag{5.19}$$

the relation (5.18) is transformed into

$$\phi^n(z) = \phi(c_n z) \cdot e^{i\gamma_n z}. \tag{5.20}$$

This functional equation can be solved completely and the solution becomes:

$$\phi(z) = \exp\left[i\delta z - \gamma|z|^{\alpha} \cdot \left\{1 + i\beta \frac{z}{|z|}\,\omega(z, \alpha)\right\}\right] \tag{5.21}$$

where

$$\omega(z, \alpha) \equiv \begin{cases} \tan\dfrac{\pi\alpha}{2}, & \alpha \neq 1, \\ \dfrac{2}{\pi}\log|z|, & \alpha = 1. \end{cases} \tag{5.22}$$

Here, α, β, γ and δ are parameters which satisfy $0 < \alpha \leq 2$, $-1 < \beta < 1$ and $\gamma > 0$. The most important of the four is α, the *characteristic exponent*. It has been proved that the normalisation factor c_n in (5.18) must be $n^{1/\alpha}$. The case $\alpha = 2$ corresponds to the Gaussian distribution. For $\alpha > 2$, the distribution would have negative probability, which is inconsistent. The parameter β governs symmetry of the distribution; for example, $\beta = 0$ corresponds to a symmetric distribution. δ is the parameter which translates the distribution and γ dominates the scale of X. The latter two parameters are not essential since they do not change the shape of distributions. If we disregard such parameters, characteristic functions of the stable distributions, except in the case $\alpha = 1$, can be transformed into the following simple form with two parameters:

$$\phi(z) = \exp\{-|z|^{\alpha} \cdot e^{\pm i\pi\theta/2}\}, \tag{5.23}$$

where the symbol $\pm$ takes the sign of z. Here θ is the symmetry parameter instead of β and its domain is restricted in the following region:

$$|\theta| \leq \begin{cases} \alpha, & 0 < \alpha < 1, \\ 2 - \alpha, & 1 < \alpha < 2. \end{cases} \tag{5.24}$$

See Figure 5.4.

The probability density $p(X; \alpha, \theta)$ is the Fourier transform of (5.23) and is given by

$$p(X; \alpha, \theta) = \frac{1}{\pi}\,\mathrm{Re}\int_0^{\infty} dz \exp\left\{-iXz - z^{\alpha} \cdot e^{i\pi/2\,\theta}\right\}, \tag{5.25}$$

where Re denotes the real part. From this expression it is obvious that

$$p(X; \alpha, \theta) = p(-X; \alpha, -\theta), \tag{5.26}$$

and for $\theta = 0$,

$$p(X; \alpha, 0) = p(-X; \alpha, 0). \tag{5.27}$$

That is, the distribution is symmetric.

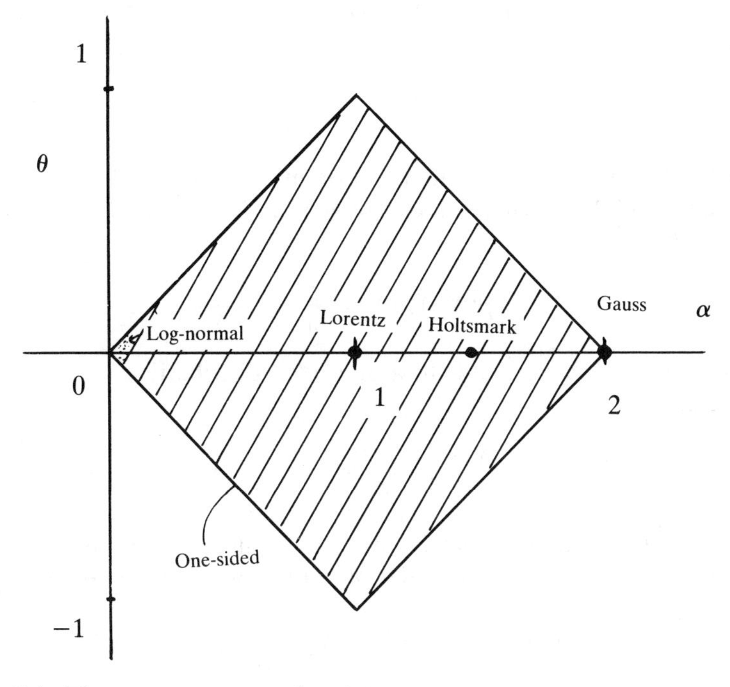

Figure 5.4 The parameter space (α, θ) for stable distribution.

Expanding the right-hand side of (5.23) in powers of X and integrating each term with respect to z, we obtain the following expansion formulae.

For $X > 0, \quad 0 < \alpha < 1$,

$$p(X;\, \alpha,\, \theta) = \frac{1}{\pi} \sum_{n=1}^{\infty} \frac{(-1)^n\ \Gamma(n\alpha+1)}{n!\ X^{n\alpha+1}} \sin \frac{n\pi}{2} (\theta - \alpha), \tag{5.28}$$

For $X > 0, \quad 1 < \alpha < 2$,

$$p(X;\, \alpha,\, \theta) = \frac{1}{\pi X} \sum_{n=1}^{\infty} \frac{(-X)^n\ \Gamma(n\alpha^{-1}+1)}{n!} \sin \frac{n\pi}{2\alpha} (\theta - \alpha). \tag{5.29}$$

Similar expressions for $X < 0$ are easily found by using (5.26). When $0 < \alpha < 1$ and $\theta = -\alpha$, we can see from (5.26) and (5.28) that for $X < 0$,

$$p(X;\, \alpha,\, -\alpha) = 0. \tag{5.30}$$

This means that the distribution is one-sided, that is, the random variable never takes negative values.

In this way, stable distributions are generally expressed in the integral form (5.25) or in the form of power series expansion as in (5.28) and (5.29). However, only four cases have been found to be expressed by elementary functions: $\alpha = 2$, $\theta = 0$ corresponds to the Gaussian distribution

$$p(X;\,2,\,0) = \frac{1}{\sqrt{\pi}}\,\mathrm{e}^{-X^2} \tag{5.31}$$

$\alpha = 1$, $\theta = 0$ corresponds to the Lorentzian (or Cauchy) distribution. (Equation (5.23) excludes the case $\alpha = 1$, but (5.25) is also valid for $\alpha = 1$, $\theta = 1$.)

$$p(X;\,1,\,0) = \frac{1}{\pi}\,\frac{1}{1 + X^2} \tag{5.32}$$

$\alpha = \frac{1}{2}$, $\theta = -\frac{1}{2}$ corresponds to a one-sided distribution:

$$p(X;\,\tfrac{1}{2},\,-\tfrac{1}{2}) = \begin{cases} 0, & X \leq 0, \\ \dfrac{1}{\sqrt{2\pi}}\,\mathrm{e}^{-1/2X} \cdot X^{-3/2}, & X > 0. \end{cases} \tag{5.33}$$

For $\alpha \doteqdot 0$, it is known that for not too small and not too large X, the distribution of X is approximated by a log-normal distribution [86]:

$$p(X;\,\alpha,\,-\alpha) \propto \frac{1}{X}\,\exp\left\{-\frac{1}{2}\,\alpha^2\,(\log X)^2\right\}. \tag{5.34}$$

The following two relations are sometimes useful in applications of stable distributions:

(1) For $X > 0$ and $\frac{1}{2} < \alpha < 1$,

$$\frac{1}{X^{\alpha+1}} \cdot p\left(\frac{1}{X^{\alpha}} : \frac{1}{\alpha},\,\gamma\right) = p(X;\,\alpha,\,\alpha(\gamma + 1) - 1). \tag{5.35}$$

(2) Let p be a variable with a stable positive distribution with index α (i.e. $p(Y;\,\alpha,\,-\alpha)$) and let X be a variable with a stable distribution with index β. Then $Z \equiv X \cdot Y^{1/\alpha}$ becomes a stable distribution with index $\alpha\beta$. For example, when we take Y as $p(Y;\,\frac{1}{2},\,-\frac{1}{2})$ and X as Gaussian, then $Z = X \cdot Y^2$ becomes Lorentzian.

The most important peculiarity of stable distributions is that, as long as $\alpha \neq 2$, they have long tails of power type for large $|X|$, such as

$$p(X;\,\alpha,\,\theta) \propto X^{-\alpha-1}. \tag{5.36}$$

Note that, for asymmetric distributions, the longer tail (in the positive or negative direction) obeys this relation but the other side decays faster than this. The existence of such tails is expected from the fact that the characteristic function given by (5.23) is singular at the origin. Since the characteristic function is the Fourier transform of its distribution function, its singularity at the origin corresponds to the singularity of the distribution function at infinity. Using the cumulative distribution function, the power law distribution (5.36) is written as

$$P(X;\, \alpha,\, \theta) \equiv \int_{X}^{\infty} p(X';\, \alpha,\, \theta)\, \mathrm{d}x' \propto X^{-\alpha}, \tag{5.37}$$

and this is the type of Equation (1.24). Therefore, in the case that X is a quantity expressing some kind of length, the characteristic exponent α can be regarded as the fractal dimension. At this point, stable distributions cannot be separated from fractals.

If the distribution has a power-type long tail as in (5.37), the qth-order absolute moment $\langle |X|^q \rangle$ diverges. Hence, for $\alpha \neq 2$, the variance is always divergent:

$$\langle X^2 \rangle = \infty. \tag{5.38}$$

Among stable distributions, the Gaussian distribution is very exceptional and its moment is finite for any order.

Historically, stable distributions other than the Gaussian have not attracted much of the attention of physicists. This seems to be due to the divergence of variance. Many physicists are apt to think that divergent variance is unphysical. However, even if the variance of a population of some distribution is infinite, this does not mean that infinity should be observable. For a finite number of samples, the observed variance is finite with probability unity just as in the case with finite variance. However, as the number of samples is increased, the variance becomes larger and larger showing a tendency to diverge. This is not unphysical at all – in fact, there are many experimental examples. Every fractal distribution is such a case, but the most appropriate example would be the $1/f$ noise described in Section 2.5. The variance of voltage fluctuation in an electric circuit which produces the $1/f$ noise increases as the period of measurement is increased. This tendency persists until the technical limit of observation is reached. In such a case it is natural to regard the variance as divergent.

The well-known central limit theorem states that a quantity made of a number of stochastic variables follows a Gaussian distribution. As a result of this theorem, the Gaussian distribution is treated specially in every field of natural science and is frequently applied. However, the fact that this theorem has a limitation is not so well known. For a counter-example, any sum of random variables of a stable distribution follows, by definition, a stable distribution with the same characteristic exponent. Hence, if its characteristic exponent is less than 2, the sum never approaches the Gaussian. The following generalised central limit theorem clarifies the conditions for the ordinary central limit theorem to hold.

For random variables $X_1, X_2, \ldots, X_n$, let $Y_n = \sum_{i=1}^{n} X_i$. If the distribution of Y_n, with an appropriate normalisation, converges to some distribution R in the limit $n \to \infty$, then the distribution R is stable. In particular, when its variance is finite, R is the stable distribution with characteristic exponent 2, that is, the Gaussian distribution.

It is clear from this theorem that the ordinary central limit theorem is only applicable to random variables with finite variance and that stable distributions other than the Gaussian are fundamental for stochastic phenomena with divergent variance.

While a number of phenomena with fractal distribution have been found recently, no one has succeeded in explaining from a general point of view why their distributions follow the power law of Equation (1.25). The above 'generalised central limit theorem' is expected to play an important role in this problem. According to this theorem, a sum of many random variables with divergent variances is likely to follow a stable distribution which has a power tail. So, if we can decompose a phenomenon into superposition of elementary random processes whose variances are divergent, then we may expect a power tail of a stable distribution to be observed.

Now we shall consider some simple physical models of stable distributions in addition to the models of turbulence of Section 4.1 and the relaxation of polymers of Section 2.4.

Models of Lorentzian distribution

(1) When vectors (w, u) on a plane are distributed isotropically, the distribution of the ratio of their components $x = u/w$ is Lorentzian.

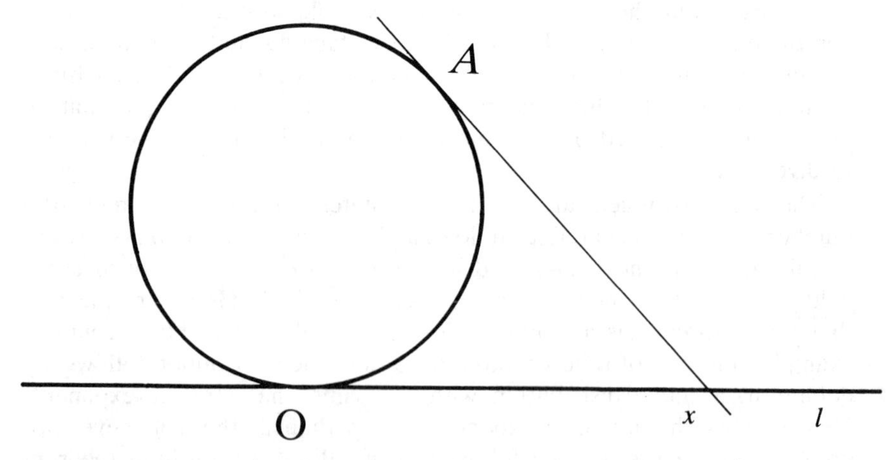

Figure 5.5 The point A uniformly distributes on the circle; x and l follows a Lorentzian distribution.

In particular, the ratio of two independent Gaussian variables obeys a Lorentzian distribution.

(2) Consider a circle and its tangent l. Choose a point A on the circle with uniform probability, then let x be the position of intersection of the tangent at point A and the line l. The distribution of x on l is Lorentzian (see Fig. 5.5).

(3) Consider the situation that Brownian particles on a plane are emitted from the origin and stick to a fixed line not passing through the origin. The distribution of the position x on the line of the attached particles is Lorentzian.

Model of a one-sided stable distribution

Let us observe a Brownian particle on a line through a slit as shown in Figure 5.6. If the particle is emitting light, we get a sequence of light pulses. The distribution of time intervals between two succeeding light pulses becomes the one-sided stable distribution, $p(x; \frac{1}{2}, -\frac{1}{2})$.

Stochastic processes whose displacements follow independent stable distributions are called *stable processes*. For example, the Lévy flight introduced in Section 1.4. is approximately a stable process, because the displacement has a power law like (1.24) and the sum of many such displacements asymptotically approaches a stable distribution. The

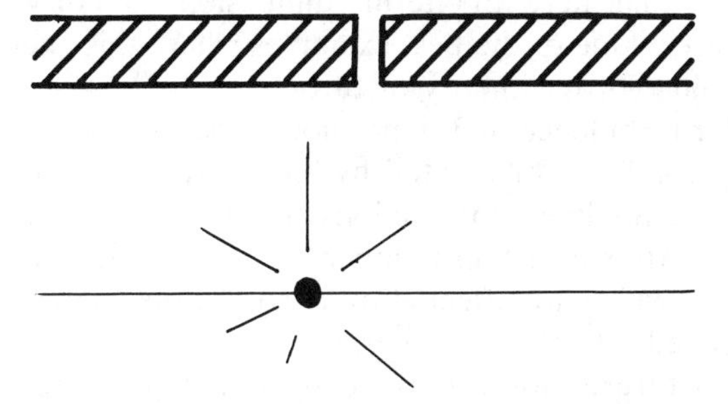

Figure 5.6 One-sided stable distribution is obtained if we observe a random walker through a slit.

stochastic equation describing such a random walk is not a usual diffusion equation. For displacements with symmetric stable distribution of characteristic exponent α, the probability density $p(x,t)$ of a particle emitted at $x=0$ satisfies

$$\hat{p}\,(k,\,t) = \int_{-\infty}^{\infty} \mathrm{e}^{ikx}\,p(x,\,t)\,\mathrm{d}x \propto \exp(-ct|k|^{\alpha}). \tag{5.39}$$

Therefore $\hat{p}(k,\,t)$ is governed by the differential equation

$$\frac{\partial}{\partial t}\,\hat{p}(k,\,t) = -c|k|^{\alpha}\,\hat{p}(k,\,t). \tag{5.40}$$

For $\alpha = 2$, we obtain an ordinary diffusion equation, but when $\alpha \neq 2$ this equation cannot be expressed in the form of a usual differential equation because $|k|^{\alpha}$ corresponds to a differential operation of non-integral order, as will be seen in Section 5.4. The diffusion equation extended in this way is a basic equation which describes random fractal variations: for example, the temporal variation of stock prices is approximated by the above equation with $\alpha = 1.7$.

5.3 **Dimensional analysis**

In the study of a physical system, important information is obtained from so-called dimensional analysis. Dimensional analysis is based on decomposing a physical quantity into the most fundamental quantities such as length, time and mass. For example, since velocity is defined by dividing length by time, we can write

$$[\mathrm{V}] = \frac{\mathrm{L}}{\mathrm{T}}, \tag{5.41}$$

where V, L and T denote the velocity, length and time, and the brackets [. . .] are used to represent its dimension. To decompose a physical quantity into L, T and M (mass) in this way is called *dimensional analysis* in physics. (Note that here the term 'dimension' is not used in the mathematical sense.) One of the best examples where this analysis proves effective is in Kolmogorov's theory of turbulence [87].

Let us consider turbulence in 3-dimensional space. A large eddy given by the boundary condition or created by an external force is broken up into smaller ones by nonlinear interactions and, these are again broken to still smaller ones. After a certain number of such divisions there appear the smallest eddies which are affected by viscosity; their kinetic energy of rotation is dissipated into thermal energy.

Though this picture is not confirmed by rigorous mathematics, it is widely accepted and called an 'energy cascade'. From an energetical point of view, we can interpret this picture such that the kinetic energy of large-scale motion is successively transformed into that of smaller scales. Since the energy spectrum $E(k)$ represents mean kinetic energy of modes whose wave vector lies in $(k, k + dk)$, the most important thing is to know the k-dependence of $E(k)$.

In turbulence with a very high Reynolds number, it is expected that there should be no characteristic length scale, as was mentioned before. That is, in the intermediate-scale region called the inertial range, the energy spectrum may be independent of boundary conditions and of viscosity. The only quantity which can affect the energy spectrum is the energy dissipation rate, denoted by ε. In a stationary state which is realised by continuous injection of kinetic energy, this quantity is equal to the injected energy rate per unit mass, and also equal to the energy flux through the inertial range.

If we assume that $E(k)$ is determined by ε only, dimensional analysis leads us to the following scaling solution of $E(k)$. Since $E(k)$ is the kinetic energy spectrum per unit mass which satisfies

$$\tfrac{1}{2}\,V^2 = \int_0^\infty E(k)\,dk, \tag{5.42}$$

the dimension of $E(k)$ is determined as

$$[E(k)] = \left(\frac{L}{T}\right)^2 \cdot L = \frac{L^3}{T^2}, \tag{5.43}$$

where we have used $[k] = 1/L$.

On the other hand, because ε is the energy dissipation rate, we have

$$[\varepsilon] = \left[\frac{d}{dt}\left(\frac{V^2}{2}\right)\right] = \left(\frac{L}{T}\right)^2 \cdot \frac{1}{T} = \frac{L^2}{T^3}. \tag{5.44}$$

If we assume the following functional form for $E(k)$:

$$E(k) \propto k^a \cdot \varepsilon^b, \tag{5.45}$$

then the dimensions of both sides become

$$\frac{L^3}{T^2} = L^{-a} \cdot \left(\frac{L^2}{T^3}\right)^b = \frac{L^{-a+2b}}{T^{3b}}, \tag{5.46}$$

which gives

$$3 = -a + 2b, \quad 2 = 3b, \tag{5.47}$$

or $a = -5/3$ and $b = 2/3$. Thus we obtain

$$E(k) \propto k^{-5/3} \cdot \varepsilon^{2/3}, \tag{5.48}$$

This is *Kolmogorov's −5/3 law*.

In the above discussion, we have implicitly assumed that ε is spatially homogeneous and not fractal. When the energy dissipation region forms a fractal structure with dimension D, we cannot neglect the k-dependence of ε. Let us imagine the fluid divided into cubes of length $1/k$. The total number of cubes is, of course, proportional to k^3. Among these cubes, there are some active cubes which belong to the energy dissipation region. The number of such cubes is estimated as

$$N\left(\frac{1}{k}\right) \propto \left(\frac{1}{k}\right)^{-D} = k^D. \tag{5.49}$$

Hence the probability that a given cube is active, $p(1/k)$, becomes

$$P\left(\frac{1}{k}\right) = k^{D-3}, \ (k > 1). \tag{5.50}$$

If we assume that $\varepsilon = 0$ outside the energy dissipation region, the value of the energy dissipation rate in active cubes of size $1/k$ is given as

$$\varepsilon^* = \langle \varepsilon \rangle \cdot p\left(\frac{1}{k}\right)^{-1}, \tag{5.51}$$

where $\langle \varepsilon \rangle$ is the mean energy dissipation rate averaged over all space. Using (5.50) and (5.51) we have

$$\begin{aligned} \langle \varepsilon^{2/3} \rangle &= (\varepsilon^*)^{2/3} \cdot p\left(\frac{1}{k}\right) \\ &= \langle \varepsilon \rangle^{2/3} \cdot k^{\frac{D-3}{3}}, \end{aligned} \tag{5.52}$$

and we finally obtain

$$\begin{aligned} E(k) &\propto k^{-5/3} \cdot \langle \varepsilon \rangle^{2/3} \cdot k^{\frac{D-3}{3}} \\ &= \langle \varepsilon \rangle^{2/3} \cdot k^{-\frac{5}{3}-\frac{3-D}{3}}. \end{aligned} \tag{5.53}$$

The last term in the exponent is the correction term caused by the fractal effect [88].

An experimental observation of the energy spectrum $E(k)$ is plotted in log-log scale in Figure 5.7 [89]. This supports the estimation of Kolmogorov's dimensional analysis. The straight part in the inertial range approximately follows $k^{-5/3}$, but the accuracy of measurement is not high enough to check the correction term in (5.53).

Another good example of dimensional analysis is in Flory's theory of polymers [90]. Consider a chain polymer which consists of N_s monomers, each of radius a. This polymer is assumed to be flexible and bent randomly by thermal fluctuation. If it spreads over a radius R, then the mean density of monomers within this range, ρ_c, is given by

$$\rho_c = \frac{N_s}{R^d}, \tag{5.54}$$

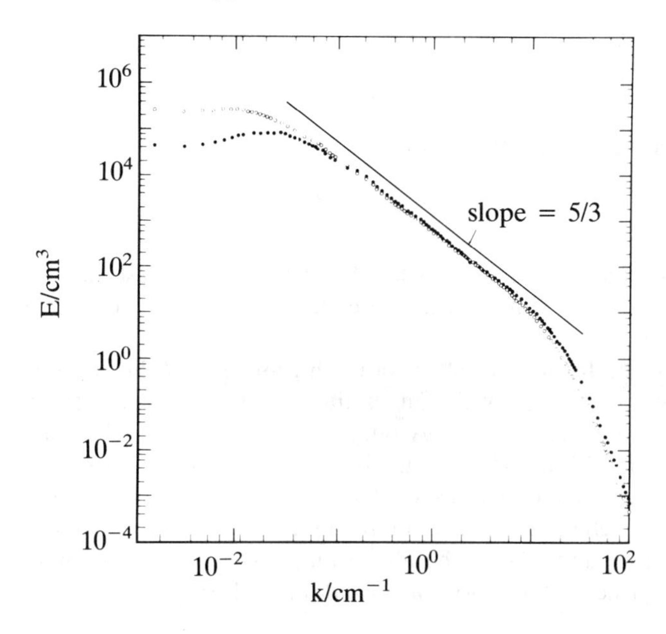

Figure 5.7 Energy spectra of a three-dimensional turbulent flow.

where d is the spatial dimension. Applying the mean-field approximation, the free energy F of this chain is estimated as

$$F \propto a^d \, R^d \, \rho_c^2 + \frac{R^2}{N_s a^2}. \tag{5.55}$$

The first term represents a repulsive force due to the excluded volume effect, which tends to make the polymer linear. The second term comes from the fluctuation, which prevents the polymer from extending. The structure of the polymer is determined by a balance between these two terms. Here we assume that the polymer has a fractal structure of dimension D, that is,

$$N_s \propto R^D. \tag{5.56}$$

Substituting (5.54) and (5.56) into (5.55), the repulsive force term is estimated as

$$a^d \, R^d \, \rho_c{}^2 \propto R^{2D-d}, \tag{5.57}$$

and the fluctuation term as

$$\frac{R^2}{N_s a^2} \propto R^{2-D}. \tag{5.58}$$

In order to balance these two terms, their exponents must be identical:

$$2D - d = 2 - D. \tag{5.59}$$

This equation gives the fractal dimension D as

$$D = \frac{2 + d}{3}. \tag{5.60}$$

In 3-dimensional space, the fractal dimension is thus estimated as 5/3, which is in good agreement with the experimental value given in Section 2.4.

Note that the formula (5.60) is not valid for large d. For d larger than 4, the correct value of D is 2. This limitation on d is derived from (5.57) and (5.58). Since the free energy ought to increase if R is increased, the exponents in (5.57) and (5.58) must be positive, namely $2D - d \geq 0$ and $2 - D \geq 0$. These inequalities lead to $D \leq 2$ and $d \leq 4$. The random structure of a chain polymer can be modelled by the trajectory of a self-avoiding random walk. The above limitation is closely related to the fact that the fractal dimension of a random walk trajectory is at most 2.

5.4 **Fractional calculus**

The order of differentiation or integration can be extended to fractional values. We can deal with a quantity such as the derivative of order 0.24 or the 0.35th-order integral. This generalisation is called fractional calculus and has a close relation to fractals.

For a sine wave e^{ikx}, applying n times the differential operator d/dx is equivalent to multiplying by $(ik)^n$,

$$\frac{d^n}{dx^n} e^{ikn} = (ik)^n \cdot e^{ikx}. \tag{5.61}$$

Where n is a natural number, this equality is trivial. For a negative integer n, (5.61) holds if we define the nth order derivative by the $|n|$th order integral. (Here we neglect the terms arising from integral constants.) For non-integral n, (5.61) can be viewed as the definition of differentiation or integration of non-integral order. In this way, it is easy to consider differentiation or integration of non-integral order for sine waves.

Since any function can be decomposed into a superposition of sine functions and the operations of integration and differentiation are linear, we may define the nth-order derivative of a given function $f(x)$ through the Fourier transformation as

$$\frac{d^n}{dx^n} f(x) \equiv \frac{1}{2\pi} \int_{-\infty}^{\infty} (-ik)^n \hat{f}(k) \cdot e^{ikx} \, dk. \tag{5.62}$$

The case $n = 0$ corresponds to the ordinary Fourier transformation. There, $\hat{f}(k)$ is computed by the inverse transformation. For integral n this formula agrees with that of usual differentiation or integration.

Consider a function $f(x)$ whose spectrum $S(k)$ is given in power of k as:

$$S(k) \equiv |\hat{f}(k)|^2 \propto k^{-\alpha}. \tag{5.63}$$

It has been already mentioned in Section 1.3 that this exponent α and the fractal dimension of its graph, D, satisfy the following relation:

$$\alpha = 5 - 2D, \tag{5.64}$$

for $1 < \alpha < 3$. The spectrum of the nth derivative $S^{(n)}(k)$ becomes, from (5.62),

$$S^{(n)}(k) \propto k^{-\alpha+2n} \tag{5.65}$$

Using the above relations (5.63)–(5.65), we obtain the fractal dimension of the graph of $f^{(n)}(x)$, as

$$D^{(n)} = D + n. \tag{5.66}$$

This relation is valid for non-integral n as long as n lies between 1 and 2. It is interesting that the fractal dimension of the graph can be changed continuously by changing the order of integration. For any positive n we have $D^{(n)} > D$ which means that differentiation increases the fractal dimension of the graph, that is, the graph becomes less smooth than the original one. Conversely for negative n we have $D^{(n)} < D$ and integration makes the graph smoother.

The definition (5.62) may be intuitively acceptable, but it involves mathematical problems such as those of convergence. A mathematically rigorous definition of non-integral-order differentiation or integration is given as follows [91]:

Let $\varphi(x)$ be integrable in any finite interval and assume that it goes rapidly to zero as $x \to \infty$, then the integral

$$\begin{aligned} I^n\varphi(x) &\equiv \frac{1}{\Gamma(n)} \int_{-\infty}^{x} \varphi(y)\,(x-y)^{n-1}\,dy \\ &= \frac{1}{\Gamma(n)} \int_{0}^{\infty} \varphi(x-y)\,y^{n-1}\,dy, \end{aligned} \tag{5.67}$$

exists for any positive number n and is called nth-order integral of $\varphi(x)$. If φ is N times continuously differentiable, I^n is defined as

$$I^n\varphi(x) \equiv \frac{1}{\Gamma(n)}\, P \int_{0}^{\infty} \varphi\,(x-y)\,y^{n-1}\,dy, \tag{5.68}$$

for $-N < n < 0$, where P denotes the principal part of a divergent integral. I^n defined as above has the following properties:

$$I^{\lambda}\,(I^{\mu}\,\varphi) = I^{\lambda+\mu}\,\varphi, \quad I^0\varphi = \varphi, \tag{5.69}$$

$$I^{-r}\,\varphi(x) = \frac{\mathrm{d}^r}{\mathrm{d}x^r}\,\varphi(x), \quad (r = 0, 1, 2, \ldots, N). \tag{5.70}$$

If we choose a pcwer or exponential function for φ, we have

$$I^{\lambda}(x^{\mu}) = \frac{\Gamma(\mu + 1)}{\Gamma(\lambda + \mu + 1)}\,x^{\lambda+\mu}, \quad x \geq 0, \tag{5.71}$$

$$I^{\lambda}(\mathrm{e}^{ax}) = \frac{1}{a^{\lambda}}\,\mathrm{e}^{ax}, \quad a > 0. \tag{5.72}$$

From these equations, we find that I^n is really a generalisation of nth-order integration. Equations (5.63)–(5.66) also hold with this difinition of fractional calculus.

The usual Brownian motion $B(x)$ is obtained by integrating white noise. The spectrum of white noise contains all frequencies and is proportional to k^0. Therefore the spectrum of $B(x)$ is proportional to k^{-2}. As derived from (5.65), the fractal dimension of the graph of $B(x)$ is 1.5. By integrating this Brownian motion a fractional number of times, we can produce a new random motion $B_H(x)$, $(0 < H < 1)$:

$$\begin{aligned} B_H(x) &\equiv I^{H-1/2}\,(B(x)) \\ &= \frac{1}{\Gamma(H + \frac{1}{2})}\int_{-\infty}^{x} (x - y)^{H-1/2}\,\mathrm{d}B(x). \end{aligned} \tag{5.73}$$

This integral diverges in general; however, the difference $B_H(x + X) - B_H(x)$ is finite, and therefore, meaningful. The case $H = 1/2$ corresponds to the original Brownian motion. For $H > 1/2$, $B_H(x)$ is a stochastic process which is less rough than Brownian motion. H is related to the fractal dimension D by

$$D = 2 - H. \tag{5.74}$$

$B_H(x)$ has self-similarity in the sense that

$$B_H(x + X) - B_H(x) \stackrel{\mathrm{d}}{=} h^{-H}\,\{B_H(x + hX) - B_H(x)\} \tag{5.75}$$

where h is an arbitrary positive quantity. That is, the variation between distant places has a similar distribution to that between near places. The beautiful pictures in Mandelbrot's book [1] of random reliefs similar to the surface of the Earth are based on this fractional Brownian motion $B_H(x)$ generalized to $(2 + 1)$-dimensional space. When x is a 2-dimensional vector, the graph of $B_H(x)$ has a shape that reminds us of a real relief on the Earth, provided we choose H approximately. This method is very important especially in application to computer graphics.

For more detailed mathematical properties of this random process, see Section 6.4.

5.5 Mean field approximation

The mean field approximation [92] is a general method which reduces a complicated many-body problem to a one-body problem in a self-consistent way. Although it is a rough approximation, we can extract some properties of the many-body system with very simple calculations. This method is very different from the renormalisation group method, so we can view the phase transition phenomena and fractals from another angle. Furthermore, it has recently been discovered that the mean field approximation and its generalised versions give quantitatively good results if we combine the approximations in a subtle way called the *coherent anomaly method*. Let us take the percolation problem by way of example.

Consider a percolation problem on a 2-dimensional square lattice with sites occupied randomly with probability p. We calculate here the fraction of sites occupied by the infinite cluster, p_∞. This quantity is equal to the probability that the origin belongs to the infinite cluster. In order that the origin belongs to the infinite cluster, the origin must be occupied and at least one of the four neighbours must belong to the infinite cluster (see Fig. 5.8). Hence we have the following equation:

$$p_\infty = p\{1 - (1 - p_\infty)^4\}. \tag{5.76}$$

Here we have assumed that the neighbours are independent of each other. This assumption, called the mean field assumption, turns the

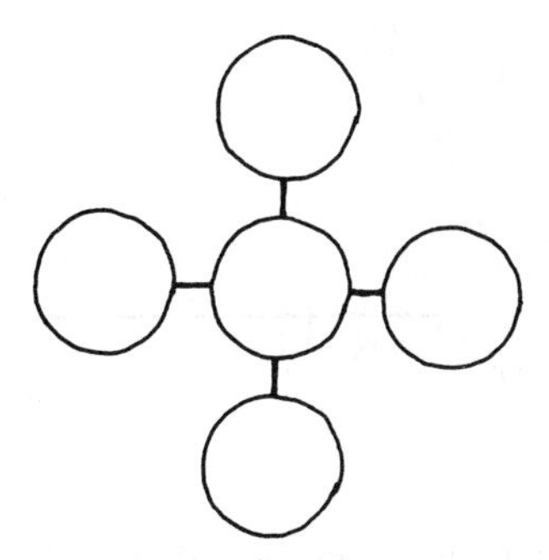

Figure 5.8 Mean field approximation for percolation.

many-body problem of percolation into a simple one-body nonlinear problem.

Equation (5.76) is an algebraic equation for p_∞ and can be solved easily. In the vicinity of the critical point, higher-order terms in p_∞ can be neglected since they are very small. Then we have the solution as

$$p_\infty = 0 \tag{5.77}$$

or

$$p_\infty = \frac{2}{3p}\left(p - \frac{1}{4}\right). \tag{5.78}$$

Obviously, there is a branching in the solution. The probability that the origin belongs to the infinite cluster becomes finite when $p > \frac{1}{4}$ (see Fig. 5.9). Thus, the percolation phase transition occurs at $p_c = \frac{1}{4}$. It may be a surprise that the phase transition can be deduced from such a simple discussion.

Qualitatively, this method is excellent; however, there are quantitative problems. As seen from (5.78), the critical exponent, β, becomes 1. This is not good compared with the known value β = 0.14. Also, the critical value $p_c = \frac{1}{4}$ is not close to the value 0.59.

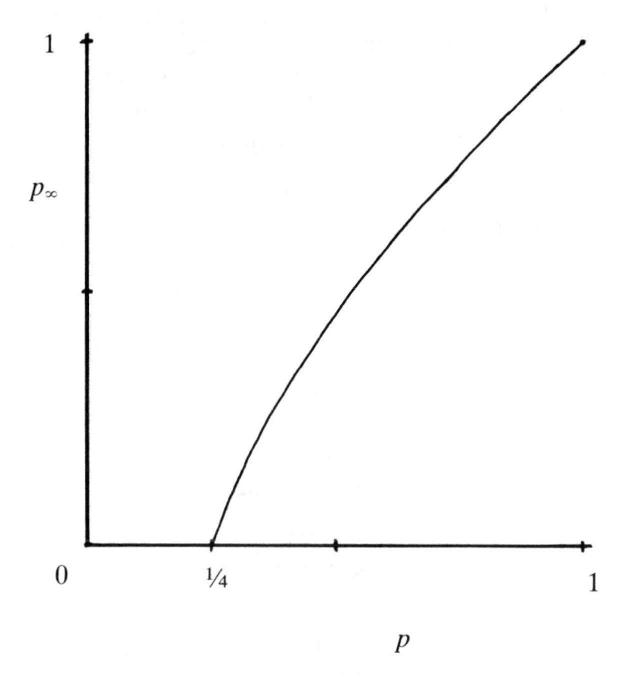

Figure 5.9 Fraction of infinite cluster plotted against p. Phase transition occurs at $p = 1/4$.

We can extend this method to higher-order approximations. In the next approximation, we calculate all configurations of nearest neighbours and replace the second-nearest neighbours by mean fields. In their case (shown in Fig. 5.10) we have the following equation:

$$p_\infty = p[1 - p\,(1 - (1 - p_\infty)^2)^4], \tag{5.79}$$

with the solutions:

$$p_\infty = 0 \tag{5.80}$$

or

$$p_\infty = \frac{2\left(p + \frac{1}{\sqrt{8}}\right)}{6p^3 + p^2}\left(p - \frac{1}{\sqrt{8}}\right). \tag{5.81}$$

Hence $p_c = 1/\sqrt{8}$ and $\beta = 1$. We have a better p_c, but the same β. In the kth approximation, we can deduce an equation for p_∞ by considering all configurations within distance k and replacing the peripheral sites by mean fields. The resulting critical value $p_c(k)$ is expected to converge to the real critical value as $k \to \infty$, but the critical exponent β does not change. This difficulty had been considered as limiting the usefulness of the approximation.

Recently, Suzuki [93] has broken through the difficulty. He pays attention to the change of the coefficients of the solutions (5.78) and

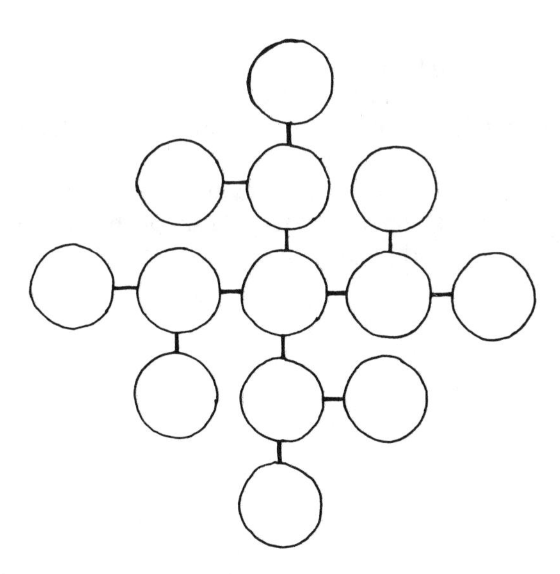

Figure 5.10 A higher-order approximation.

(5.81). Denoting the kth approximation by $p_\infty = a(k)(p - p_c(k))$, the following relation is assumed:

$$a(k) \propto (p_c^* - p_c(k))^{-\Psi} \tag{5.82}$$

where p_c^* and $p_c(k)$ denote the critical probability of the exact and the kth approximation, respectively. This assumption is based on the intuition that as $p_c(k)$ comes close to the exact value p_c^*, the coefficient will diverge because it diverges in the real solution.

Now p_∞ is expressed as

$$p_\infty \propto (p_c^* - p_c(k))^{-\Psi} \cdot (p - p_c(k)). \tag{5.83}$$

If we put $p \doteqdot p_c^*$, then we have

$$p_\infty \propto (p_c^* - p_c(k))^{1-\Psi}. \tag{5.84}$$

This relation implies

$$\beta = 1 - \Psi. \tag{5.85}$$

Therefore if we can estimate Ψ, we have a revised value for β. With more than 3 approximations, we can actually determine β and p^*_c from (5.82) [94]. With two approximations and the known p^*_c, we have the critical exponent by the following formula.

$$\beta = 1 - \Psi = 1 + \frac{\log \dfrac{a_2}{a_1}}{\log \dfrac{p_c^* - p_c(2)}{p_c^* - p_c(1)}} . \tag{5.86}$$

Even with the approximations $k = 1$ and $k = 2$, we have $\beta = 0.16$, which is a great improvement over the mean field value 1. Of course this method can be applied to other critical exponents, including the fractal dimension of percolation clusters.

As seen from this example we may say that the mean field approximation with this coherent anomaly method is a simple but very powerful tool for complicated many-body problems.

6 Notes on fractal dimension

This chapter can be regarded as an appendix. Its purpose is to arrive at a deeper understanding of the fractal dimension from some specific problems and mathematical topics.

6.1 Extensions of fractal dimension

The fractal dimension is the most basic quantity for the representation of self-similar shapes and phenomena. But, it is impossible to describe more complicated shapes with only one number. We need to extend the fractal dimension. There are two approaches to the extension. One is to generalise the fractal dimension so as to depend on the scale of observation. By this generalisation, it becomes possible to define the dimension of a shape as a function of length including upper and lower cut-off scales. The other approach is to introduce new quantities to describe spatial fluctuations of fractal dimensions. In this way, we will cope with objects the local fractal dimensions of which change from part to part.

As described in Chapter 1, any fractal object in nature has a lower and upper cut-off length, and self-similarity can be found only in the limited scale of observation in between these lengths. Of course, the fractal dimension is meaningful only in that scale range. There have been attempts to extend fractal dimension to scales where self-similarity is not realised. Considering the fact that the fractal dimension is a very powerful tool for describing complicated shapes, it is desirable that we extend the range of applications. Mandelbrot suggested one possibility of such extension by introducing an 'effective dimension' [1]. His idea is as follows:

Consider a ball of 10 cm diameter made of a thick thread of 1 mm diameter ... To an observer placed far away, the ball appears as a zero-dimensional figure: a point ... As seen from a distance of 10 cm resolution, the ball of thread is a three-dimensional figure. At 10 mm, it is a mess of one-dimensional threads. At 0.1 mm, each thread becomes a column and the whole becomes a three-dimensional figure again. At 0.01 mm, each column dissolves into fibres, and the ball again becomes one-dimensional, and so on, with the dimension crossing over repeatedly from one value to another. When the ball is represented by a finite number of atomlike pinpoints, it becomes zero-dimensional again.

This example shows that the effective dimension of an object depends on the scale of observation. We can discuss this intuitive idea in a more rigorous way as follows.

We first review the definition of the fractal dimension. Let r denote the size of coarse-graining and $N(r)$ the number of unit shapes, for example, spheres. Then the fractal dimension is defined as

$$D = -\frac{\log N(r)}{\log r}, \tag{6.1}$$

if the right-hand side is independent of r. This dimension cannot always be defined, since the right-hand side of (6.1) is constant only when $N(r)$ follows an exact power law. We want to extend the fractal dimension to more general functions $N(r)$. The fractal dimension D in (6.1) can be viewed as the slope of the graph of $N(r)$ in log-log plot. Hence, it seems to be most natural to define a scale-dependent fractal dimension by the slope of the graph of $N(r)$, that is, the derivative of $\log N(r)$ with respect to $\log r$:

$$D(r) = -\frac{\mathrm{d} \log N(r)}{\mathrm{d} \log r}. \tag{6.2}$$

We can define this quantity $D(r)$ as long as $N(r)$ is a smooth function. It agrees of course with the usual fractal dimension in the case when $N(r)$ obeys a power law.

Solving (6.2) inversely, the following expression is deduced:

$$N(R) = N(r) \cdot \exp\left(-\int_r^R \frac{\mathrm{D}(s)}{s}\, \mathrm{d}s\right). \tag{6.3}$$

This expression shows how observations of different scales are linked by the scale-dependent fractal dimension.

Now, let us see how this extended fractal dimension is applied in a physical problem. Here we consider a random walk whose mean free path is finite. Though the fractal dimension for a trajectory of ideal Brownian motion is 2, in the case of finite mean free path we expect that the fractal dimension depends on observation scale. If we observe with a scale much shorter than the mean free path, we will find that the trajectory is nearly

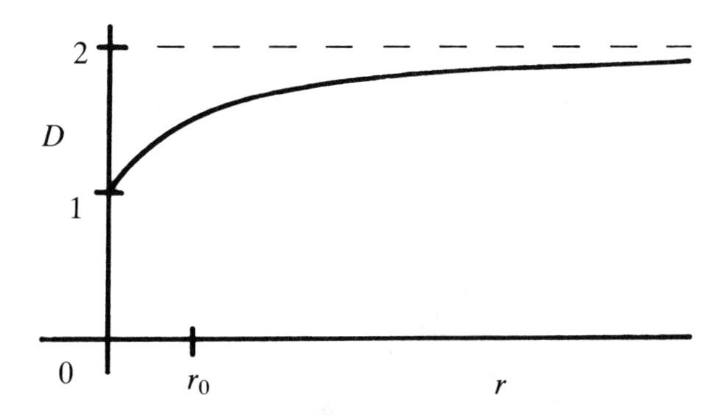

Figure 6.1 Scale-dependent fractal dimension for a random walk with finite mean free path.

a straight line. On the other hand, observing with a larger scale, the random walk may be reduced to the usual Brownian motion.

For a Markovian random walk in 1-dimensional space whose mean free path is finite, the following exact solution is obtained by using the renormalisation group method [95]:

$$D(r) = 2 - \frac{1}{1 + \dfrac{r}{r_0}}, \tag{6.4}$$

where r_0 is a parameter proportional to the mean free path. From Figure 6.1 we can find a transition of the fractal dimension from $D(0) = 1$ to $D(\infty) = 2$. This result supports the above intuitive discussions. (Here, $D(r)$ takes a greater value than the dimension of space, $d = 1$. Note that $D(r)$ is not the Hausdorff dimension but the potential dimension D^* which will be considered in Sections 6.2 and 6.4.)

From (6.3) and (6.4) we obtain the relation

$$L(r) \propto \frac{1}{1 + \dfrac{r}{r_0}}, \tag{6.5}$$

where $L(r)$ is the length of a trajectory of random walks measured by unit length r (here $L(r) = rN(r)$). It has been confirmed that this equation is valid not only for a random walk in 1-dimensional space but also for a trajectory of a particle in a hard-sphere gas in 3 dimensions [96]. In Figure 6.2, dots show averaged values of $L(r)$ for a trajectory of a particle in a numerically simulated hard-sphere gas [97], and the solid line represents the curve given by (6.5). As seen from the figure, they agree very nicely in their first two significant digits. Furthermore, it has been shown

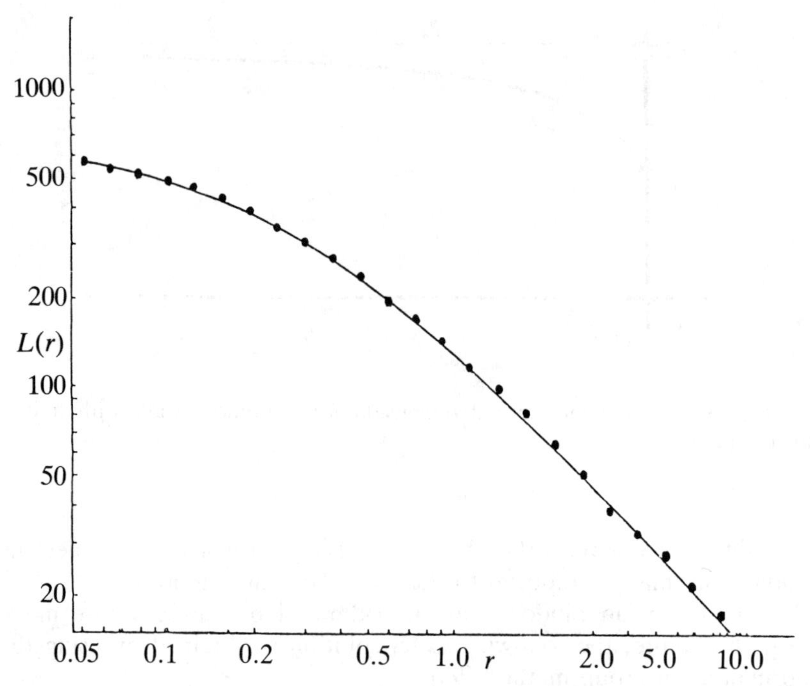

Figure 6.2 Scale-dependent length of a trajectory of a molecule in hard-sphere gas.

experimentally that (6.5) is also applicable to a real random walk of a fine particle in solution as observed through a microscope [98].

Another way of extending the fractal dimension is to introduce higher-order fractal dimensions to make up for lack of information that cannot be described by the fractal dimension alone.

For example, even if we know that the fractal dimension of a given set is, for example, 1.3, it is impossible to judge whether it is a set consisting of discrete points or of crumpled lines. Such topological information is described by a quantity called *topological dimension*, d_T. This dimension is, in a sense, more fundamental than the fractal dimension. As will be mentioned in Section 6.4, it takes an integer value, and is invariant under any continuous transformation. The topological dimension of a set is 0 if it can be transformed into a point (or a set of isolated points) by a continuous mapping. For a set which can be transformed into a line the topological dimension is 1. Thus the topological dimension is 0 for the Cantor set, and 1 for the Koch curve. For the Sierpinski gasket, it is not trivial but in fact $d_T = 1$. Generally the fractal dimension is greater than or equal to the topological dimension [1].

As mentioned in Chapter 1, the fractal dimension is a generic name for dimensions which can take fractional values. These includes the Hausdorff dimension D_H, the capacity dimension D_c, and the information dimension D_I. For some fractals, such as de Wijs's fractal, these dimensions do not coincide. This discord may seem perplexing, but from the stand-point of characterising sets in more detail it turns out to be useful. Assume that there are two sets which have the same capacity dimension but look different. We cannot distinguish these two sets by the dimension alone. However, if their information dimensions take different values, then we can classify the two sets by the information dimensions. Hence, the more higher-order fractal dimensions we have which can take different values, the better we can distinguish complicated sets or shapes. In the following we shall see a possible way to define infinite series of higher-order fractal dimensions [99].

Suppose that points are distributed randomly in a d-dimensional space. Divide the space into d-dimensional cubes of side ε. Let P_i denote the probability that a point belongs to the ith cube. For an arbitrary positive number q, we introduce a new quantity $I_q(\varepsilon)$ given by the following equation:

$$I_q(\varepsilon) = \frac{1}{1-q} \log \sum_i P_i^q. \tag{6.6}$$

We define the *qth-order fractal dimension* as

$$D_q = \lim_{\varepsilon \to 0} \frac{I_q(\varepsilon)}{\log (1/\varepsilon)}. \tag{6.7}$$

The quantity I_q is often called the *qth-order Renyi information* and we can easily show that I_1 coincides with the usual information.

$$\begin{aligned} I_1(\varepsilon) &= \lim_{q \to 1} \frac{1}{1-q} \log \sum_i P_i^q \\ &= \lim_{\delta \to 0} \left\{ -\frac{1}{\delta} \log \left(1 + \delta \sum_i P_i \log P_i\right) \right] = -\sum_i P_i \log P_i. \end{aligned} \tag{6.8}$$

As a consequence D_1 coincides with the information dimension D_I. The dimension D_q is also called the qth-order information dimension.

Consider the limit $q \to +0$; we can show that D_0 agrees with the capacity dimension, D_c. From the following equation:

$$\lim_{q \to +0} P_i^q = \begin{cases} 0, & P_i = 0, \\ 1, & P_i \neq 0, \end{cases} \tag{6.9}$$

it is not difficult to understand that $\lim_{q \to 0} \Sigma P_i^0$, is equal to $N(\varepsilon)$, the number of cubes which include at least one point. Hence (6.7) becomes identical to (1.4), and $D_0 = D_c$.

For an integer q, it is possible to give a physical meaning to D_q as follows. Assume that there are M points $\{x_i\}$ in a Euclidean space and consider clusters which contain q points whose distances apart are all less than ε. Let $N_q(\varepsilon)$ denote the number of such groups. The qth-order correlation integral is defined by

$$C_q(\varepsilon) = \lim_{M \to \infty} M^{-q} \cdot N_q(\varepsilon). \tag{6.10}$$

It is known that $C_q(\varepsilon)$ is approximately equal to $\exp\{(1 - q)\ I_q(\varepsilon)\}$. Hence assuming the behaviour of $C_q(\varepsilon)$ near $\varepsilon = 0$ as, $C_q(\varepsilon) \propto \varepsilon^{\tau_q}$, D_q is linked to τ_q by the following equation:

$$\tau_q = (q - 1)\ D_q, \tag{6.11}$$

where the index τ_q is called the *qth-order correlation exponent*. When we compute the fractal dimension of a strange attractor of chaos, $\tau_2 = D_2$ is a very useful quantity. This exponent indicates how the number of pairs whose distances are less than ε changes with ε. It is obvious that this quantity is equivalent to the fractal dimension introduced on p. 17 by the correlation function method.

There are two important properties of D_q. One is that D_q is a decreasing function of q, that is, the following inequality is satisfied:

$$D_q \geq D_{q'} \quad (q < q'). \tag{6.12}$$

From this inequality, we have the following relation as a special case:

$$D_c \geq D_I \geq \tau_2. \tag{6.13}$$

As stated in Chapter 1, the Hausdorff dimension D_H satisfies $D_c \geq D_H \geq D_I$. So, in general, the qth-order information dimension D_q is not larger than D_H. In the case that points are distributed uniformly in $\mathbb{R}^d$, we have $D_q = d = D_H$ for all $q \geq 0$.

The other important property is that D_q is invariant under diffeomorphism, or differentiable mappings. Any dimension except for the topological dimension, D_T, can change under a non-smooth deformation of space such as a non-differentiable map. But, for smooth deformation D_q can be proved to be invariant. We will argue this point in detail in Section 6.4.

Recently, this generalised information dimension has been attracting great attention in connection with multi-fractal or $f - \alpha$ spectra [100]. The $f - \alpha$ spectrum is introduced as follows in order to characterise non-uniformity of fractal distribution: let $P_i(\varepsilon)$ be the probability of finding a particle in ith box of size ε. In the limit $\varepsilon \to 0$, we assume p_i to behave as

$$P_i(\varepsilon) \propto \varepsilon^{\alpha_i}. \tag{6.14}$$

Roughly speaking α_i denotes the local fractal dimension for the probability distribution at the location i. We consider the case that α_i differs from place to place. Let the probability of α_i taking a value between α and $\alpha+d\alpha$ be scaled by ε as

$$n(\alpha)\ \varepsilon^{-f(\alpha)}\ d\alpha. \tag{6.15}$$

Here $f(\alpha)$ corresponds to the fractal dimension of the set where α_i is equal to α. From (6.15) we can replace the sum in the logarithm in (6.6) by the following integral:

$$\sum_i P_i^q = \int d\alpha\ n(\alpha)\ \varepsilon^{-f(\alpha)+q\alpha}. \tag{6.16}$$

In the limit $\varepsilon \to 0$, the integrand shows a sharp peak at $\alpha = \alpha_q$ where α_q makes the exponent $-f(\alpha) + q\alpha$ smallest. Evaluating the integral by the saddle-point-approximation, we obtain the relations:

$$D_q = \frac{1}{q-1}\,[q\alpha_q - f(\alpha_q)], \tag{6.17}$$

$$f(\alpha_q) = q\alpha_q - (q-1)\ D_q, \tag{6.18}$$

$$\alpha_q = \frac{d}{dq}\,[(q-1)\ D_q]. \tag{6.19}$$

The function $f(\alpha_q)$ is called the *$f - \alpha$ spectrum* and its diagram is called the *$f - \alpha$ diagram*.

From (6.17) the maximum of $f(\alpha)$ is always equal to D_0. If we draw the tangent to the curve of $f(\alpha)$ from the origin in the $f - \alpha$ diagram, its slope is 1 in any case and it attaches at $f(\alpha) = \alpha = D_1$ (information dimension). For ideal fractal with perfect similarity such as the Koch curve, the f-α diagram becomes a single point (see also section 6.4.9).

We can calculate $f(\alpha)$ exactly for de Wijs's fractal introduced in section 1.4. Let b denote the division rate of de Wijs's fractal instead of the α in Section 1.4, to avoid confusion. The probability distribution of this fractal is obtained by binomial coefficients and it is easy to calculate D_q:

$$\begin{aligned} D_q &= \lim_{n\to\infty} \frac{1}{q-1}\,\frac{\log \Sigma\ {}_nC_m\ \{b^m(1-b)^{n-m}\}^q}{\log 2^{-n}} \\ &= -\frac{1}{q-1}\log_2 \{b^q + (1-b)^q\}. \end{aligned} \tag{6.20}$$

Then we have,

$$\alpha_q = -\frac{b^q \log_2 b + (1-b)^q \log_2 (1-b)}{b^q + (1-b)^q}, \tag{6.21}$$

$$f(\alpha_q) = q\ \alpha_q + \log_2 (b^q + (1-b)^q). \tag{6.22}$$

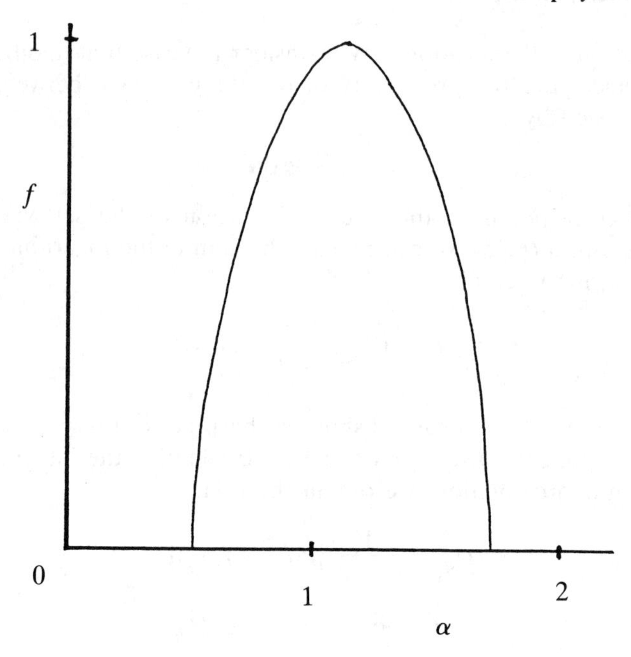

Figure 6.3 f-α spectrum of the de Wijs fractal.

The corresponding $f - \alpha$ diagram is shown in Figure 6.3 for the case $b = 0.3$. Thus the de Wijs fractal is characterised by a convex curve, which shows quantitatively the non-uniformity of this fractal.[1]

So far we have defined the dimension of sets in Euclidean space using the usual root square distance, $r = \sqrt{(x_1^2 + x_2^2 + \ldots + x_n^2)}$. It is not necessarily appropriate for all problems. For example, let us consider percolation on a square lattice. In this case, it is sometimes physically more meaningful to measure the distance between two points by the minimal number of bonds which connect the two points. For example, in the problem of metal–insulator percolation the motion of an electron is restricted along connected bonds. Dealing with such a problem, a basic quantity is the number of distinct sites which an electron can reach within n time steps. When $N(n)$ increases as a power of n, we know the structure of connected sites possesses a fractal property. Let us introduce a new dimension in order to specify it:

$$N(n) \propto n^{\hat{D}}. \tag{6.23}$$

This dimension is called the *spreading dimension* or connectivity dimension [101]. Obviously $\hat{D}$ is not affected by the embedding space. So, it is possible that the spreading dimension of a structure on a plane exceed 2,

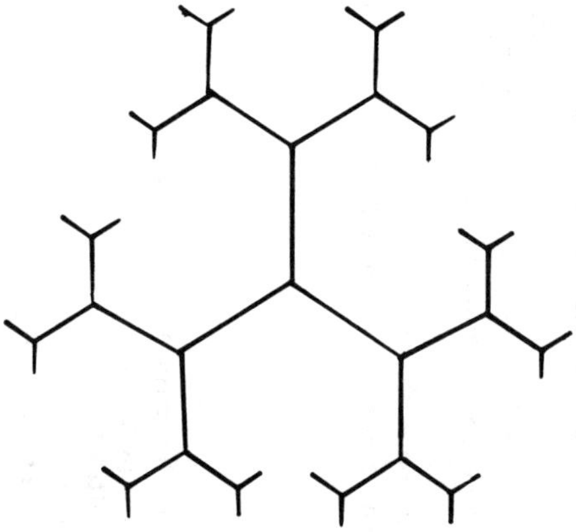

Figure 6.4 A Cayley tree.

the dimension of the plane. For example, in the case of the Cayley tree as shown in Figure 6.4, the spreading dimension becomes infinite.

6.2 Summary of various dimensions

Throughout the book we have introduced many fractal dimensions, but have not paid much attention to the differences between them in order to avoid inessential arguments for readers who are not familiar with non-integral dimensions. However, for readers who like rigorous arguments, the practical and rough application of fractal dimensions may have been confusing. Hence, we collect here the different dimensions and discuss the relations between them. The various notations are reviewed in Table 6.1. All non-integral dimensions, except for the spectral dimension and spreading dimension, are greater than d_{T} and less then d.

$$d_{\mathrm{T}} \leq D,\ D_{\mathrm{s}},\ D_{\mathrm{H}},\ D_{\mathrm{c}},\ D_{\mathrm{I}},\ D_q,\ D_{\mathrm{L}} \leq d. \quad (q > 0) \tag{6.24}$$

In particular, in Euclidean space d_{T} equals d, so all these dimensions become identical.

Further, the following inequalities are known:

$$D_{\mathrm{c}} \geq D_{\mathrm{H}} \geq D_{\mathrm{I}} \geq D_q \quad (q > 1), \tag{6.25}$$

$$D_{\mathrm{s}} \geq D_{\mathrm{H}}. \tag{6.26}$$

Relating to (6.26), an interesting example in which D_{s} is not equal to D_{H} is known. Consider the set of rational numbers in the interval [0, 1]. As is well known this set is countable, hence we have $D_{\mathrm{H}} = 0$. (Note that any countable set of points can be covered by a countable number of spheres

Table 6.1 Summary of dimensions

Symbol	Name	Description	Page
d	Euclidean dimension	Dimension of the Euclidean space in which the observed set is embedded; takes only integer values.	6
d_T	Topological dimension	Takes integral values only; is invariant under homeomorphism.	157
D	Fractal dimension	Generic terminology without strict definition; often identified with Hausdorff dimension or capacity dimension.	11
D_s	Similarity dimension	Defined for strictly self-similar objects.	6
D_H	Hausdorff dimension	Defined by most efficient covering.	8
D_c	Capacity dimension	Defined by covering with identical spheres or cubes.	10
D_I	Information dimension	Calculated from a probability distribution.	14
D_q	qth-order information dimension	An extension of the information dimension; capacity dimension, information dimension and correlation exponent are included as special cases.	146
D_L	Lyapunov dimension	Used to characterise the dimension of the chaotic attractor.	66
$\tilde{D}$	Spectral dimension	Related to random walk properties on fractal structures.	102
$\hat{D}$	Spreading dimension	Not resticted by the embedded space; shows a fractal connectivity.	148
$D(r)$	Scale-dependent dimension	Generalised as a function of observation length	142

of infinitesimal rad., therefore the D-dimensional Hausdorff measure is zero for any positive D.) On the other hand, the set of rational numbers is self-similar and its similarity dimension is 1, which is also equal to the capacity dimension. In general, when D_s is defined,

$$D_c = D_s \tag{6.27}$$

is expected to hold.

Further, the following equation is also expected to hold for the closure of any self-similar set:

$$D_c = D_s = D_H. \tag{6.28}$$

In physics, it is sufficient to consider only closed sets, so in practice (6.28) is always true.

For a 'measure-theoretic support', that is, a subset on which the measure of the considered set is concentrated, the Hausdorff dimension D'_H is known to be equal to the information dimension:

$$D'_H = D_I. \tag{6.29}$$

In a technical sense the following relations are expected to hold for strange attractors of chaos[2]

$$D_c = D'_c \geq D_I = D_L = D'_H \geq D_2, \tag{6.30}$$

where D'_c and D'_H represent capacity and Hausdorff dimensions of the 'measure-theoretic support' of the attractor. Numerical research suggests that the difference between D_c and D_L is about 2–10%.

As regards spectral dimension, the following relation is obtained for the Sierpinski gasket and for its higher-dimensional versions:

$$D_H \geq \widetilde{D}. \tag{6.31}$$

According to Mandelbrot [102], $\widetilde{D}$ can be expressed by other fractal dimensions as follows:

$$\widetilde{D} = \frac{2D_H}{D_w} = 2(1 - D_R), \tag{6.32}$$

where D_H, D_w and D_R denote Hausdorff dimensions of the considered fractal structure, the dimension of the trajectory of random walk and the dimension of the set of recurrence times on the time axis, respectively. Here, D_R is allowed to be negative when the random walker does not return to the starting point. In such a case, from (6.32) we have $\widetilde{D} > 2$. Conversely, for the case $\widetilde{D} < 2$, $D_R > 0$ is obtained, which means that the random walk is recurrent. It is important to recognise that the spectral dimension can be thus reduced to more fundamental fractal dimensions.

The spreading dimension $\hat{D}$ is determined from connectivity between points, and is independent of the embedding space except for the space with $d_T = 1$. So, in general, $\hat{D}$ has no relation with other dimensions. However, if we consider structures on a Euclidean lattice in $\mathbb{R}^d$, the following relation is expected to hold:

$$D_2 \geqslant \hat{D}. \qquad (6.33)$$

This is because the ordinary distance is always smaller than the distance defined on a connecting bond.

We shall now explain the important quantity, potential dimension, denoted by D^*, which has already appeared in several places. We know that the Hausdorff dimension of the trajectory of Brownian motion equals 2 in a space with dimension larger than 2. It is often convenient to regard the dimension of a trajectory as 2 even in 1-dimensional space. We call such a quantity *potential dimension*.

Potential dimension can be either negative or greater than d, the dimension of the space. The relation to Hausdorff dimension is given as follows:

$$D_H = \begin{cases} 0, & D^* \leq 0, \\ D^*, & 0 < D^* < d, \\ \mathrm{d}, & D^* \geq d. \end{cases} \qquad (6.35)$$

As will be described in Section 6.4, it is possible to consider direct products, projection and intersections of fractal sets. By using the potential dimension we can calculate the Hausdorff dimension of such sets in a general way.

There are still other definitions of fractal dimensions. Readers who are interested in this topic are recommended to read references [1] and [55].

6.3 **Methods of analysing sequential data**

The temporal fluctuation of a quantity $\{x(t)\}$ is easily obtained by experiment or observation. We describe here how to extract fractal properties from such fluctuations.

Generally, it is possible to classify any random temporal fluctuation into two categories, stationary and non-stationary. Here, stationary means that the n-fold distribution function $p(x(t_0), x(t_0 + t_1), \ldots, x(t_0 + t_n))$ does not depend on the origin of the time-axis. We can tell whether a given fluctuation is stationary or not from its power spectrum.

Let $S_x(t)$ denote the power spectrum of $x(t)$ and let us assume it follows a power law in the vicinity of $f = 0$:

$$S_x(f) \propto f^{-\gamma}. \qquad (6.36)$$

As described in Section 1.3, when the exponent γ lies on $(0, 1)$, this quantity is proportional to the Fourier transform of the correlation function:

$$\langle x(t)\, x(t+\tau)\rangle \propto \begin{cases} \tau^{\gamma-1} & (0 < \gamma < 1), \\ \delta(\tau) & (\gamma = 0). \end{cases} \tag{6.37}$$

This quantity is independent of t, which shows that the fluctuation is stationary. Actually in this case, $\{x(t)\}$ is stationary, and its fractal dimension on the time-axis is given by γ. In particular, in the case $\gamma = 0$, the fluctuation becomes white noise. From the standpoint of fractals we can also call it 0-dimensional noise. For $\gamma > 1$, (6.37) is not valid and the fluctuation is non-stationary. In any non-stationary data, absolute time measured from a fixed origin is meaningful as well as relative time, so it becomes very hard to analyse it. The $1/f$ noise described in Section 2.5, for which $\gamma = 1$, is located at the borderline between stationary and non-stationary.

From any signal of the form (6.37), we can produce a signal whose exponent lies in the interval $(0, 2)$. Consider the nth derivative of the function $x(t)$. As described in Section 5.4, for the spectrum of the nth derivative of $x(t)$, we have the relation:

$$S_x(n) \propto f^{-\gamma+2n}. \tag{6.38}$$

The fractal dimension D of the derivative is given by the following if the exponent lies between -1 and 0:

$$D = \gamma - 2n. \tag{6.39}$$

If the given signal looks like a set of pulses rather than a continuous variation, it is sometimes convenient to investigate the distribution of intervals between pulses. First, decide on a threshold x_c as shown in Figure 6.5. Then consider a step-like function $f_{x_c}(t)$ which is 1 when $x(t)$ is greater than x_c and is 0 in other cases. We have a sequence of pulses and intervals. Obtain the distribution of the intervals. If the cumulative distribution of intervals $P(\tau)$ satisfies the following relation, then the distribution of pulses can be said to be D-dimensional.

$$P(\tau) \propto \tau^{-D}. \tag{6.40}$$

Fractal properties of temporal fluctuation can also be found in the distribution of displacements: the fluctuation of stock prices introduced in Section 2.5 is a good example. The distribution of displacement follows a stable law with characteristic exponent 1.7. In that case we examine the distribution of displacement Δx in a fixed time interval Δt, that is, the distribution of $x(j\Delta t) - x((j-1)\Delta t)$. If this distribution decays exponentially as $|\Delta x|$ tends to infinity for a fixed Δt, then the distribution is

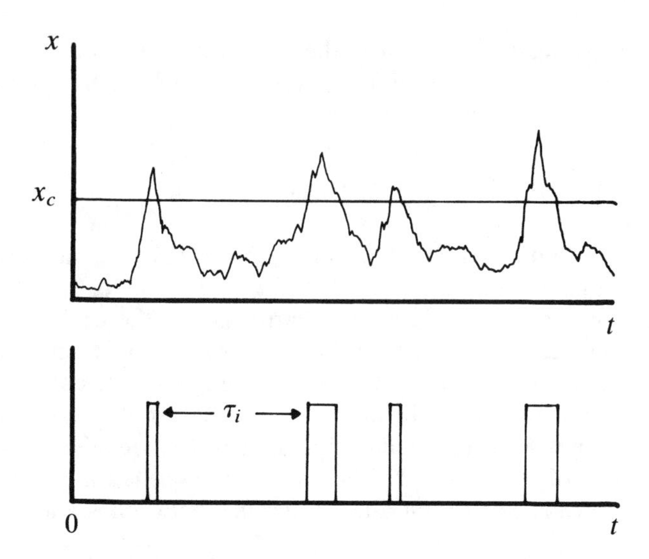

Figure 6.5 Transform of a continuous fluctuation into pulses.

expected to gradually approach a Gaussian as Δt becomes larger. The distribution of displacement is not fractal in that case. When the distribution has a fractal property, its higher-order moment diverges. Hence, if there are some mechanisms which suppress the divergence of the fluctuation, then the fluctuation cannot have a fractal property.

Random fluctuation is sometimes caused by a strange attractor in higher-dimensional space. In such a case, the fractal dimension of the underlying attractor might be guessed from the data of the time sequence $\{x(t)\}$. It is not simple to obtain information about an attractor from a time sequence. Many researchers have tried to develop a practical method since Takens mathematically proved the possibility in 1981 [103]. Among other ideas we here introduce the method proposed by Sano and Sawada [104]. From the set of data $\{x(t)\}$ we construct an orbit in d-dimensional 'delay coordinate' as follows:

$$\vec{X}(t) \equiv (x(t) \cdot x(t + t_d), x(t + 2t_d), \ldots, x(t + (d - 1)t_d)), \qquad (6.41)$$

where $t_d > 0$ denotes the delay time. Arbitrarily choose one point $\vec{X}(t_0)$ on the orbit. Find the points on the orbit whose distance from $\vec{X}(t_0)$ is less than ε, and denote them by $\{\vec{X}(t_j); j = 1, 2, \ldots, m\}$. If there is no such point, change the centre point $\vec{X}(t_0)$ to another one and find the $\{\vec{X}(t_j)\}$ for the new centre. Next, consider the displacement vector for each point:

$$\vec{Z}^j(\tau) \equiv \vec{X}(t_j + \tau) - \vec{X}(t_0 + \tau), \quad j = 1, 2, \ldots, m. \qquad (6.42)$$

If ε is sufficiently small, $\vec{Z}^j(\tau)$ can be expressed by a linear transformation as follows:

$$\vec{Z}^j(\tau) = A(\tau) \cdot \vec{Z}^j(0), \quad j = 1, 2, \ldots, m, \tag{6.43}$$

where $A(\tau)$ is a $d \times d$ matrix. There are various methods of determining $A(\tau)$ from the data $\{\vec{Z}^j(0)\}$ and $\{\vec{Z}^j(\tau)\}$ for a fixed τ. It is most natural to determine $A(\tau)$ by the least-squares method, which minimises the sum of the squared norm $\Sigma\ |\vec{Z}^j(\tau) - A(\tau)\vec{Z}^j(0)|^2$. It has been confirmed that $A(\tau)$ satisfies the following equation:

$$A(\tau) \cdot V = C, \quad (V)_{kl} \equiv \frac{1}{m} \sum_{j=1}^{n} Z^j_k(0)\ Z^j_l(0),$$

$$(C)_{kl} \equiv \frac{1}{m} \sum_{j=1}^{m} Z^j_k(\tau)\ Z^j_l(0), \tag{6.44}$$

where $(\quad)_{kl}$ represents the (k,l) component of the $d \times d$ matrix, and Z_k represents the kth component of $\vec{Z}$. The matrix $A(\tau)$ is uniquely determined by (6.44) if $m \geq d$ and it is not degenerate. The Lyapunov exponent can be calculated from $A(\tau)$ as

$$\lambda_j = \langle \frac{1}{\tau} \log |A(\tau)\ \vec{e}_j| \rangle, \tag{6.45}$$

where $\langle\cdot\rangle$ represents the ensemble average over different $\vec{X}(t_0)$, and $\{\vec{e}_j;\ i = 1, 2, \ldots, d\}$ denotes the set of orthogonal unit vectors. It has been confirmed that Lyapunov exponents obtained by this method are generally in good agreement with known values in some standard dynamical systems. This refers not only to positive values but also to zero and even negative ones.

Using these Lyapunov exponents the fractal dimension D of the attractor is given by the following equation as described in Section 3.2:

$$D = j - \frac{\sum_{i=1}^{j} \lambda_i}{\lambda_j} \tag{6.46}$$

where $\lambda_1 \geq \lambda_2 \geq \ldots \geq \lambda_d$, $j = \min\ \{n\ |\lambda_1 + \lambda_2 + \ldots + \lambda_n < 0\}$. If D is constant for different d (dimension of delay coordinate) and τ (delay time), we may conclude that the dimension of the attractor is D. This means that the fluctuation $\{x(t)\}$ is a projection of a motion on a D-dimensional attractor in a higher-dimensional space, namely its degree of freedom is finite.[3] The problem remains of deciding how effective this method is for various random fluctuations in nature. Perhaps in some cases D is not constant but increases with d. In such a case it is concluded that the number of degrees of freedom is infinite.

6.4 **Mathematical back-up**

In this section, we introduce some mathematical results in order to deepen our understanding of fractals. Each of these subsections is independent of the others, so they may be read separately.

6.4.1 *How to calculate the Hausdorff dimension*

It is generally very complicated to calculate the Hausdorff dimension rigorously. As a simple example we illustrate here how to determine the Hausdorff dimension of a Cantor set [105].

First, let us define a set of closed intervals E_j called a 'net': $E_0 = [0,1]$, $E_1=[0,\frac{1}{3}] \cup [\frac{2}{3},1]$, $E_2 = [0,\frac{1}{9}] \cup [\frac{2}{9},\frac{1}{3}] \cup [\frac{2}{3},\frac{7}{9}] \cup [\frac{8}{9},1]$ .... Thus E_{j+1} is defined as the set of intervals produced by eliminating the central third of each interval E_j. The interval E_j consists of 2^j intervals with length 3^{-j}. The Cantor set is expressed as $E = \cap_{j=0}^{\infty} E_j$. If D = log2/log3, then the D-dimensional Hausdorff measure of E satisfies

$$M_D(E) \leq \lim_{j\to\infty} 2^j \cdot (3^{-j})^D = 1. \tag{6.47}$$

because E is covered by an ensemble of E_j.

Next, let C be the set of closed intervals. We arbitrarily choose a closed interval I from C. Then consider the elements of the net in I. Any two closed intervals in the net can take only two states: either the two intervals are separated from each other, or one includes the other, so we can cover all points of E in I by two separated closed intervals of the net, J and J'. (Here J and J' are not necessarily included in the same E_j.) Let $K \subset I$ be an interval which does not contain any points of E and which lies between J and J'. Due to the property of the net, the length of K is given by $|K|=\max(|J|, |J'|)$, where $|\cdot|$ denotes the length of an interval. Hence the following inequality holds:

$$\begin{aligned} |I|^D &\geq (|J| + |K| + |J'|)^D \\ &\geq (\frac{3}{2} (|J| + |J'|)^D \\ &= 2(\frac{1}{2}|J| + \frac{1}{2}|J'|)^D \geq |J|^D + |J'|^D. \end{aligned} \tag{6.48}$$

That is, if we decompose I into J and J' the covering becomes more efficient. By repeating such decomposition by nets, any coverings of E can be reduced to E_j in which the length of every interval is 3^{-j}. Hence,

$$\sum_{I\in c} |I|^D \leq \sum_{J\in E_j} |J|^D = 1. \tag{6.49}$$

By definition $M_D(E)$ is the lowest value of the left-hand side of this inequality, so we get the following relation:

$$M_D(E) \geq 1. \tag{6.50}$$

Therefore we have

$$M_D(E) = 1. \tag{6.51}$$

From the definition of Hausdorff dimension, the D-dimensional Hausdorff measure of E can be finite only when D equals the dimension of E. Hence we finally obtain $D_{\mathrm{H}} = D = \log 2/\log 3$ for the Cantor set.

6.4.2 *Scale invariance and power functions*

In mathematical expressions fractals are always described by power functions. This is for the following reason. A function $f(x)$ is scale-invariant if $f(x)$ is proportional to the scaled function $f(\lambda x)$ for all λ. That is, if $f(x)$ is scale-invariant then there exists a function $C(\lambda)$ such that

$$f(x) = C(\lambda)\, f(\lambda x) \tag{6.52}$$

Differentiating both sides with respect to x and eliminating $C(\lambda)$ we have

$$f'(x)/f(x) = \lambda\, f'(\lambda x)/f(\lambda x). \tag{6.53}$$

Substituting $x = 1$ and integrating both sides with respect to λ, the solution of this functional equation is obtained as

$$f(x) = f(1)\, x^{\alpha}, \quad \alpha = f'(1)/f(1). \tag{6.54}$$

Thus we know that there is no scale-invariant function other than the power functions.

6.4.3 *A curious property of the Cantor set*

The Lebesgue measure of the Cantor set is 0 but it contains uncountably many points. Thus the Cantor set has the following curious property. Let x_1 and x_2 be points in a Cantor set. It has been proved that the distance between the two points, $|x_1 - x_2|$, can take any value from 0 to 1 if we choose suitable x_1 and x_2 [105]. In 2-dimensional space, this implies that a set $(E \times [0,1] \cup ([0,1] \times E)$ contains any rectangle whose sides have length less than 1. Thus it is possible to find any size of rectangles in Figure 6.6.

6.4.4 *Definition of topological dimension*

There are several ways to define topological dimension d_{T}. Here, we introduce an inductive definition.

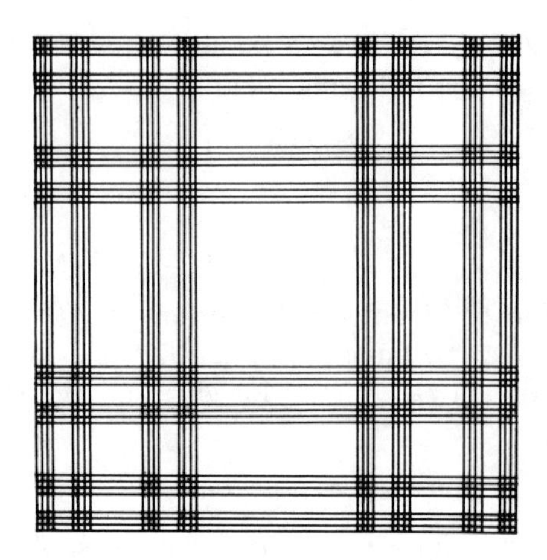

Figure 6.6 A strange Cantor mesh. There are rectangles of every size.

First, we define $d_T(\phi) \equiv -1$ for the empty set ϕ. Assume that for an integer n, we have d_T defined for any set R with $d_T(R) \leq n - 1$. Then we define the dimension of R with $d_T(R) \leq n$ by the following. For given R, there exists an open set V which satisfies $d_T(\bar{V} - V) \leq n - 1$ and $F \subset V \subset G$, where F and G are arbitrary open and closed sets in R, respectively, and satisfy $F \subset G$. (Note that $\bar{V}$ denotes the closure of V.) Thus we can define $d_T(R) \leq n$ assuming $d_T(R) \leq n - 1$. Now we define $d_T(R) = n$ for a set R where R satisfies $d_T(R) \leq n$ and does not satisfy $d_T(R) \leq n - 1$. Thus by using mathematical induction, d_T is defined for any natural number. Intuitively, the definition is constructed by the inductive method from lower dimension by using the fact that the boundary of an n-dimensional set, $\bar{V} - V$, is an $(n - 1)$-dimensional set.

6.4.5 *Invariance of D_q under diffeomorphism*

The generalised information dimension D_q has been proved to be invariant under any differentiable 1-to-1 transformations, $f: \vec{x} \to \vec{x}$ [99]. The proof is as follows.

A d-dimensional cube of side length ε centred at $\vec{x}$ is transformed to a distorted shape $A(\vec{x}\,')$ by the transformation f. We denote by $\varepsilon_0(A(\vec{x}))$ the side length of the minimum cube which includes $A(\vec{x})$, and by $\varepsilon_1(A(\vec{x}))$ the side length of the maximum cube included by $A(\vec{x})$. There exist constants C_1, $C_2 > 0$ such that for any $\vec{x}$

$$C_1 \cdot \varepsilon \leq \varepsilon_1 \leq \varepsilon_0 \leq C_2 \cdot \varepsilon, \tag{6.55}$$

where ε is a sufficiently small constant. As q-th-order information $I_q(\varepsilon)$ is a decreasing function for ε, the following relations hold

$$I'_q\,(C_2 \cdot \varepsilon) \leq I_q\,(\varepsilon) \leq I'_q\,(C_1 \cdot \varepsilon), \tag{6.56}$$

where $I'_q(\varepsilon)$ is the qth-order information defined in the distorted space by dividing the space into cubes of size ε. From the above relations, we have the following inequality in the limit $\varepsilon \to 0$:

$$D'_q \leq D_q \leq D'_q. \tag{6.57}$$

Hence the qth-order information dimensions D_q and D'_q are identical. Invariance of D_{H} is also proved in a similar way. As most physical transformations are differentiable, D_q (and therefore D_{c}, D_{I} and τ_q) may be regarded as invariant.

6.4.6 *Direct products, intersection and projection of fractal sets*

These results are all from Mandelbrot [1].

Direct products

Let S_1 and S_2 be sets in Euclidean spaces $\mathbb{R}^{d_1}$ and $\mathbb{R}^{d_2}$, respectively. They are assumed to be fractals and we denote their Hausdorff dimensions by D_{H_1} and D_{H_2}. We consider direct product of S_1 and S_2 to be defined in $\mathrm{R}^{d_1+d_2}$ and denote the Hausdorff dimension of the resulting set by $D_{H_{1+2}}$. The following inequality has been mathematically proved:

$$D_{H_{1+2}} \geq D_{H_1} + D_{H_2}. \tag{6.58}$$

Here equality holds when S_1 and S_2 are independent of each other. We should be careful on the point that equality does not always hold: for some S_1 and S_2, it is known that $D_{H_{1+2}} > 0$ but $D_{H_1} = D_{H_2} = 0$.

Intersection

Let S_1 and S_2 be fractal sets in $\mathbb{R}^d$ with Hausdorff dimension D_{H_1} and D_{H_2}, respectively. We denote the potential dimension of their intersection by $S_1 \cap S_2$. A simple sum rule is known for their co-dimensions:

$$d - D^*_{H_{1\times 2}} = d - D_{H_1} + d - D_{H_2}. \tag{6.59}$$

Thus we get the following equation:

$$D^*_{H_{1\times 2}} = D_{H_1} + D_{H_2} - d. \tag{6.60}$$

Knowing the potential dimension, we can immediately determine from (6.35) the true value of the Hausdorff dimension which lies between 0 and d. In particular, in the case where S_2 is an ordinary set with integer dimension d_2 such as a line or a plane, we have

$$D^*_{H_{1\times 2}} = D_{H_1} - (d - d_2). \tag{6.61}$$

Projection

When a D_H-dimensional fractal set S in $\mathbb{R}^d$ is projected to a d_0-dimensional Euclidean subspace, the Hausdorff dimension D_H of the projected set S' is given as

$$D_H' = \min(d_0, D_H). \tag{6.62}$$

In this case, if we denote the potential dimension of S' by $D_H^{*\prime}$, (6.62) becomes $D_H^{*\prime} = D_H$.

6.4.7 *Fractal dimension of graphs*

For a given function $f(t)$, we assume that the following inequality holds for all t and h ($0 < h \leq h_0$):

$$|f(t+h) - f(t)| \leq ch^{2-D} \tag{6.63}$$

where c and h_0 are positive constants. It has been proved that the D-dimensional Hausdorff measure of the graph $(t, f(t))$ is finite. Therefore $D_H \leq D$ holds [13].

The fractional Brownian motion, $B_H(t)$, introduced in Section 5.4 statistically satisfies the following property [1]:

$$|B_H(t+h) - B_H(t)| \propto h^H \tag{6.64}$$

where $D_H = 2 - H$.

Mandelbrot investigated the validity of some methods of measuring the fractal dimension of a graph [106]. As a result, the following two methods are found to give a valid value. One method is to determine the dimension $D_H = 2 - H$ from Equation (1.9′) by means of square box counting. The other is to determine it by using $M(r)$, the number density of points within r (Equation (1.16)). However, the dimension measured using $M(r)$ gives $D_H = 1/H$. This $1/H$ is the potential dimension of the trajectory of fractional Brownian motion, as will be mentioned later. The disagreement comes from the fact that the second method disentangles multiplicative trajectories. These results are valid only in the limit $r \to 0$. In the limit $r \to \infty$, the dimension of the graph becomes 1 by any method. Indeed, the graph of Brownian motion looks linear in the large-scale limit.

6.4.8 *Properties of fractal random walks*

The reference for the following properties is [16]. Fractional Brownian motion, $\vec{B}_H(t)$, and Lévy's stable flight, $\vec{L}_\alpha(t)$, are both typical fractal random walks. Both of them are vectors in $\mathbb{R}^d$. Each component of $\vec{B}_H(t)$ consists of independent scalar fractional Brownian motion, $B_H(t)$. Lévy flight is defined by a random walk in $\mathbb{R}^d$ such that a walker jumps to completely random directions with random step length u obeying the probability $P(u) \propto u^{-\alpha}$. The region of parameters is restricted to $0 < H \leq 1$ and $0 < \alpha < 2$. While the trajectory of $\vec{B}_H(t)$ is continuous, that of $\vec{L}_\alpha(t)$ is discontinuous (hence it is called 'flight'). Potential dimensions of these trajectories are given respectively as follows:

$$D^* = \frac{1}{H}, \quad \alpha. \tag{6.65}$$

The trajectories fill up the space in the case of $D^* \geq d$.

Potential dimensions of zero sets on the time-axis are given respectively as

$$1 - dH, \quad 1 - \frac{d}{\alpha}. \tag{6.66}$$

Here, the zero set is defined by the set of points on the time-axis at which the walker arrives at the origin of $\mathbb{R}^d$. If these values are positive, random walkers return to the origin infinitely many times. On the other hand, if the values are negative, random walkers will either not or very rarely come to the origin.

The potential dimensions of N-multiple points of trajectories are given by

$$d - N(d - \frac{1}{H}), \quad d - N\,(d - \alpha). \tag{6.67}$$

In a special case the condition that trajectories have no double-point is given as follows:

$$dH > 2, \quad \frac{d}{\alpha} > 2. \tag{6.68}$$

In the situation that time is not continuous, that is, the case when we observe only at fractal times, potential dimensions of the trajectories become

$$\frac{D_t}{H}, \quad D_t \cdot \alpha, \tag{6.69}$$

where D_t denotes the dimension of observation time.

As we have seen from (6.65) and (6.69), $1/H$ and α appear in exactly the same way. Therefore, if we represent the above results by the potential dimension of trajectories in (6.65), $D^* = 1/H = \alpha$, we can unify the above results.

6.4.9 *Recent developments in multi-fractals*

Some curious but important properties of the fractal dimensions related to the study of multi-fractals have recently been found. In section 6.1, we introduced the local fractal dimension α and the global dimension $f(\alpha)$. The quantity α is closely related to the so-called Hölder exponent α_H which is defined at a point x for a given function $P(x)$ as the upper bound of H that satisfies the following relation for any small positive ε,

$$|P(x + \varepsilon) - P(x)| \leqq c(x)\, \varepsilon^{H}, \tag{6.70}$$

where $c(x)$ is independent of ε.

In the case when $P(x)$ is an integral of a distribution $p(x)$, i.e. $P(x) = \int_{-\infty}^{x} P(y)\, dy$, then the left hand side of (6.70) can be regarded as $P_i(\varepsilon)$ in (6.14), so $\alpha \leqq \alpha_H$ holds.

For practical purposes it is convienient to fix the proportionality factor in (6.14) by a constant c and define α by the following equation:

$$P_i\,(\varepsilon) \equiv \int_{x}^{x+\varepsilon} p(y)\, dy = c\varepsilon^{\alpha}. \tag{6.71}$$

Mandelbrot [112] has shown by generalising his original idea for multi-fractals [113] that both α and f can take negative values. Therefore α and f should be considered as 'potential' dimensions like D^* in section 6.2. For $f(\alpha) < 0$ the value of α cannot be estimated from a single realisation, but we have to repeat estimates many times in order to obtain a consistent estimate.

A strange $f - \alpha$ spectrum has been found for the model of a river introduced in section 4.5. The value obtained for α is everywhere 3, hence $f = 1$. Consequently, the $f - \alpha$ diagram is composed of a single point, but the generalised dimension D_q is a curve [114]. This result casts a serious doubt on the fractal dimension defined by the mass-radius relation (1.16), because the fractal dimension by this method becomes 3 and is larger than the spatial dimension of 1. It is anticipated that this dimension is equal to $D_{-\infty}$ [115].

It seems contradictory, but the fractal dimensions (except the Hausdorff dimension and the capacity dimension) might not have the geometrical meaning of 'dimension' in a rigorous sense.

This example shows how little we know of the basic properties of the fractal dimensions. We are still far from a complete understanding of fractal dimensions.

Notes

1 Recently such multi-fractal structure has actually been found in turbulence. Maneveau and Sreenivasan [111] have confirmed experimentally that the energy dissipation rate $\varepsilon(x)$ is well characterised by D_q or $f - \alpha$.

2 Especially for strange attractors of 2-dimensional diffeomorphisms the following equations are proved [55]:

$$D_c' = D_I = D_L = D_H' = D_2.$$

3 If the detailed information of $\{\lambda_i\}$ is not necessary, the set of dimensions of an attractor, $\{D_q\}$, can be obtained more easily from (6.10) and (6.11) by counting up the number of groups of neighbouring points in the delay coordinate.

References

[1] B. B. Mandelbrot, *Fractals – form, chance and dimension* (1977), revised as *The fractal geometry of nature* (W. H. Freeman, San Francisco, 1982).
[2] M. Hirata, *Kirin-no madara* (in Japanese) (Chuokoron-sha, Tokyo, 1975).
[3] T. Musha, *The world of fluctuation* (in Japanese) (Kodan-sha, 1980); T. Musha (ed.), *Proceedings of the Symposium on 1/f fluctuations* (Tokyo Institute of Technology, Tokyo, 1977).
[4] J. T. Hack, Michigan USGS, Prof. Paper, 504B, 40.
[5] Y. Y. Kagan and L. Knopoff, *Geophys. J.*, **62**(1980), 303; M. A. Sadovskiy *et al.*, *Izvestiya, Earth Physics*, **20**(1984), 87.
[6] H. Takayasu and I. Nishikawa, *Proc. of 1st. Int. Symp. for 'Science on Form'*, S. Ishizaka (ed.), (KTK Sci. Pub., 1986), 15; M. P. Wiedeman, *Circulation Research*, **12**(1963), 375.
[7] R. Thoma, *Archiv der Entwicklungsmechanik*, **12**(1901), 352; R. A. Groat, *Federation Pr.*, **7**(1948), 45.
[8] D. R. Morse *et al.*, *Nature*, **314**(1985), 731.
[9] A. S. Szalag and D. N. Schramm, *Nature*, **314**(1985), 718; E. J. Groth and P. J. E. Peebles, *Astrophys. J.*, **217**(1977), 385.
[10] H. Mizutani, *The science of craters* (in Japanese) (Tokyo University Press, 1980).
[11] J. E. Avron and B. Simon, *Phys. Rev. Lett.*, **46**(1981), 1166.
[12] L. F. Burlaga and L. W. Klein, *Jour. Geophys. Res.*, **91**(1986), 347.
[13] D. Avnir, D. Farin and P. Pfeifer, *Nature*, **308**(1984), 261; D. Farin, A. Volpert and D. Avnir, *J. Am. Chem. Soc.*, **107**(1985), 3368.
[14] S. R. Forrest and T. A. Witten Jr., *J. Phys.*, **A12**(1979), L109.
[15] M. Matsushita *et al.*, *Phys. Rev. Lett.*, **53**(1984), 286.
[16] R. M. Brady and R. C. Ball, *Nature*, **309**(1984), 225.
[17] R. Jullien and R. Botet, *Aggregation and fractal aggregates* (World Scientific Publishing, Singapore, 1987).
[18] J. Nittman, G. Daccord and H. E. Stanley, in *Fractals in physics*, L. Pietronero and E. Tosatti (eds), (North-Holland, Amsterdam, 1986), 193.
[19] J. Nittmann, G. Daccord and H. E. Stanley, *Nature*, **314**(1985), 141; J. D. Chen and D. Wilkinson, *Phys. Rev. Lett.*, **55**(1985), 1892.
[20] L. Niemeyer, L. Pietronero and H. J. Wiesmann, *Phys. Rev. Lett.*, **52**(1984), 1033.
[21] A. A. Few, *J. Geophys. Res.*, **75**(1970), 7517.
[22] J. P. Allen *et al.*, *Biophys. J.*, **38**(1982), 299.

[23] R. F. Voss *et al.*, in *The Mathematics and physics of disordered media, proceedings 1983*, B. D. Hughes and B. W. Ninham (eds), Lecture Notes in Mathematics, 1035(Springer, Berlin, 1983), 153.
[24] M. Suzuki, *Prog. Theor. Phys.*, **69**(1983), 65.
[25] A. S. Monin and A. M. Yaglom, *Statistical fluid mechanics, mechanics of turbulence 2* (MIT Press, 1975); R. A. Antonia *et al.*, *Phys. Fluids*, **241**(1981), 554.
[26] S. Lovejoy, *Science*, **216**(1982), 185.
[27] H. G. E. Hentschel and I. Procaccia, *Phys. Rev.*, **A28**(1983), 417.
[28] J. Perrin, *Les Atomes* (Alcan, Paris, 1913). English translation: *Atoms*.
[29] E. Nelson, *Phys. Rev.*, **150**(1966), 1079.
[30] L. F. Abbot and M. B. Wise, *American J. of Physics*, **49**(1981), 37.
[31] J. T. Bendler, *J. Stat. Phys.*, **36**(1984), 625.
[32] E. W. Montroll and M. F. Shlesinger, in *The mathematics and physics of disordered media proceedings 1983*, B. D. Hughes and B. W. Ninham (eds), Lecture Notes in Mathematics, 1035(Springer, Berlin, 1983), 109.
[33] G. Williams and D. C. Watts, *Trans. Faraday Soc.*, **66**(1970), 80; M. F. Shlesinger, *J. Stat. Phys.*, **36**(1984), 639.
[34] V. N. Belykh *et al.*, *Phys. Rev.*, **B16**(1977), 4860.
[35] M. Robnik, *Phys. Lett.*, **80A**(1980), 117; *J. Phys. A Math. Gen.*, **14**(1981), 3195.
[36] M. A. Caloyannides, *J. Appl. Phys.*, **45**(1974), 307; F. N. Hooge, *Physica*, **83B**(1976), 14.
[37] A. D'Amico and P. Mazzetti (eds), *Noise in physical systems and 1/f noise – 1985* (Elsevier, Amsterdam, 1986).
[38] E. W. Montroll and M. F. Shlesinger, *J. Stat. Phys.*, **32**(1983), 209.
[39] M. F. Shlesinger and E. W. Montroll, in *The mathematics and physics of disordered media, proceedings 1983*, B. D. Hughes and N. W. Ninham (eds), Lecture Notes in Mathematics, 1035(Springer, Berlin, 1983), 130.
[40] J. Crutchfield, *Physica*, **10D**(1984), 229.
[41] T. A. Witten Jr. and L. M. Sander, *Phys. Rev. Lett.*, **47**(1981), 1400.
[42] P. Meakin, *Phys. Rev.*, **A27**(1983), 1495.
[43] M. Muthukumar, *Phys. Rev. Lett.*, **50**(1983), 839.
[44] H. Gould *et al.*, *Phys. Rev. Lett.*, **50**(1983), 686
[45] K. Kawasaki and M. Tokuyama, *Phys. Lett.*, **100A**(1984), 337; K. Honda, H. Toyoki and M. Matsushita, *J. Phys. Soc. Japan*, **55**(1986), 707.
[46] P. Meakin, *Proceedings of Int. Conf. Fragmentation, Form and Flow in Fractured Media* (1986); see also [17].
[47] R. C. Ball and R. M. Brady, *J. Phys.*, **A18**(1985), L809; P. Meakin, *Phys. Rev.*, **A33**(1986), 3371; see also [17].
[48] P. Meakin, *Phys. Rev.*, **B29**(1984), 2930; R. Jullien *et al.*, *J. Phys. Lett (France)*, **45**(1984), L2111.
[49] M. Matsushita, *J. Phys. Soc. Japan*, **54**(1985), 865.
[50] E. N. Lorenz, *J. Atmos. Sci.*, **20**(1963), 130.
[51] O. E. Rössler, *Phys. Lett.*, **50**(1983), 346.
[52] D. A. Russel *et al.*, *Phys. Rev. Lett.*, **45**(1980), 1175; P. Grassberger and I. Procaccia, *Phys. Rev. Lett.*, **50**(1983), 346.
[53] J. Kaplan and J. Yorke, in *Functional differential equations and approximation of fixed points. Proceedings 1978*, H.–O. Peitgen and H.–O. Walther, Lecture Notes in Mathematics, 730(Springer, Berlin, 1979), 228.
[54] C. Simo, *J. Stat. Phys.*, **21**(1979), 465; I. Shimada and T. Nagahsima, *Prog. Theor. Phys.*, **61**(1979), 1605.

[55] J. D. Farmer, E. Ott and J. A. Yorke, *Physica*, **7D**(1983), 153.
[56] M. J. Faigenbaum, *Phys. Lett.*, **74a**(1979), 375; *J. Stat. Phys.,* **19**(1978), 25.
[57] J. P. Gollub and S. V. Benson, *Phys. Rev. Lett.*, **41**(1978), 948; *J. Fluid Mech.*, **100**(1980), 449.
[58] T. Y. Li and J. A. Yorke, *Amer. Math. Monthly*, **2**(1975), 985.
[59] J. E. Hutchinson, *Indiana Univ. Math. Jour.*, **30**(1981), 713; M. Hata, *Japan Jour. Applied Math.*, **2**(1985), 381.
[60] D. Stauffer, *Introduction to percolation theory* (Taylor & Francis, London, 1985); *Phys. Rep.*, **54**(1979), 1.
[61] L. Reatto and E. Rastelli, *J. Phys.*, **C5**(1972), 2785.
[62] L. Pietronero and H. Wiesmann, *J. Stat. Phys.*, **36**(1984), 909.
[63] M. Matsushita *et al.*, *J. Phys. Soc. Japan*, **55**(1986), 2618.
[64] H. Takayasu, *Prog. Theor Phys.*, **74**(1985), 1343.
[65] H. Takayasu, *Phys. Rev. Lett.*, **54**(1985), 1099; H. Takayasu, in *Fractals in physics*, L. Pietronero and E. Tosatti (eds) (North-Holland, Amsterdam, 1986), 181.
[66] O. Martin, A. M. Odlyzko and S. Wolfram, *Commun. Math. Phys.*, **93**(1984), 219.
[67] S. Wolfram, *Physica*, **10D**(1984), 1.
[68] I. Procaccia, *J. Stat. Phys.*, **36**(1984), 665.
[69] H. Takayasu, *Prog. Theor. Phys.*, **72**(1984), 471.
[70] S. Chandrasekhar, *Rev. Mod. Phys.*, **1**(1943), 1.
[71] H. Takayasu, *J. Phys. Soc. Japan*, **56**(1987), 1257.
[72] H. G. E. Hentschel and I. Procaccia, *Phys. Rev.*, **A29**(1984), 1461.
[73] R. Rammal, *Phys. Rep.*, **103**(1984), 505.
[74] S. H. Liu, *Solid State Phys.*, **39**(1986), 307.
[75] R. Rammal, *J. Stat. Phys.*, **36**(1984), 547.
[76] B. J. Alder and T. E. Wainwright, *Phys. Rev. Lett.*, **18**(1967), 988.
[77] M. H. Ernst and A. Weyland, *Phys. Lett.*, **34A**(1971), 39; H. van Beijeren, *Rev. Mod. Phys.*, **54**(1982), 195.
[78] H. Takayasu and K. Hiramatsu, *Phys. Rev. Lett.*, **53**(1984), 633.
[79] P. Bak and R. Bruinsma, *Phys. Rev. Lett.*, **49**(1982), 249.
[80] J. Hubbard, *Phys. Rev.*, **B17**(1978), 494.
[81] S. H. Liu, *Phys. Rev.*, **B32**(1985), 7360; see also [74].
[82] H. Takayasu, I. Nishikawa and H. Tasaki, *Phys. Rev.*, **A37**(1988); see also [6].
[83] K. G. Wilson and J. Kogut, *Phys. Rep.*, **12**(1974), 75.
[84] T. W. Burkhardt and J. M. J. van Leeuwen (eds), *Real-space renormalization* (Springer, Heidelberg, 1982).
[85] W. Feller, *An introduction to probability theory and its applications* (Wiley, New York, 1966), vol. 2; P. Lévy, *Théorie de l'addition des variables aléatoires* (Gauthier-Villars, Paris, 1937).
[86] E. W. Montroll and J. T. Bendler, *J. Stat. Phys.*, **34**(1984), 129.
[87] A. N. Kolmogorov, *C. R. Acad. Sci. USSR*, **30**(1941), 301.
[88] B. B. Mandelbrot, in *Turbulence Seminar*, *Proceedings 1976/1977*; P. Bernard and T. Raitu, Lecture Notes in Mathematics, 615(Springer, 1977), 121.
[89] F. H. Champagne, *J. Fluid Mech.*, **86**(1978), 67.
[90] P. J. Flory, *Principles of polymer chemistry* (Cornell University Press, Ithaca, NY, 1971); P. G. De Gennes, *Scaling concepts in polymer physics* (Cornell University Press, Ithaca, NY, 1979).

[91] K. B. Oldham and J. Spanier, *The fractional calculus* (Academic Press, New York and London, 1974).
[92] P. Weiss, *J. Phys. Radium*, **6**(1907), 661; for percolation; M. Stephen, *Phys. Rev.*, **B15**(1977), 5674.
[93] M. Suzuki, *Phys. Lett.*, **116A**(1986), 375; M. Suzuki, *J. Phys. Soc. Japan*, **55**(1986), 4205.
[94] M. Takayasu and H. Takayasu, *Phys. Lett. A*, **128**(1988), 45.
[95] H. Takayasu, *J. Phys. Soc. Japan*, **51**(1982), 3057.
[96] S. Tsurumi and H. Takayasu, *Phys. Lett.*, **113A**(1986), 449.
[97] D. C. Rapaport, *Phys. Rev. Lett.*, **53**(1984), 1965.
[98] S. Matsuura, S. Tsurumi and N. Imai, *J. Chem. Phys.*, **84**(1986), 539.
[99] P. Grassberger, *Phys. Lett.*, **97A**(1983), 227; R. Benzi *et al.*, *J. Phys.*, **A17**(1984), 3521.
[100] T. C. Halsey *et al.*, Phys. Rev., **A33**(1986), 1141.
[101] R. Rammal *et al.*, *J. Phys. A. Math. Gen.*, **17**(1984), L491; N. Berker and S. Ostlund, *J. Phys.*, **C12**(1979), 4951; M. Suzuki, *Prog. Theor. Phys.*, **69**(1983), 65.
[102] B. B. Mandelbrot, *J. Stat. Phys.*, **36**(1984), 541.
[103] F. Takens, in *Value distribution theory, Proceedings 1981*, I. Laine and S. Rickman (eds), Lecture Notes in Mathematics, 989 (Springer, Berlin, 1983), 336.
[104] M. Sano and Y. Sawada, *Phys. Rev. Lett.*, **55**(1985), 1082.
[105] K. J. Falconer, *The geometry of fractal sets* (Cambridge University Press, 1985).
[106] B. B. Mandelbrot, *Physica Scripta*, **32**(1985), 257.
[107] B. B. Mandelbrot, *J. Stat. Phys.*, **34**(1984), 895.
[108] L. Pietronero and E. Tosatti (eds), *Fractals in physics* (North-Holland, Amsterdam, 1986).
[109] A. N. Kolmogorov, Entropy per unit time as a metric invariant of automorphisms, *Dokl. Akad. Nauk. SSSR*, **124**(1959), 754 (English summary in *Math. Rev.*, **21**, 2035).
[110] C. R. Doering *et al.*, *Phys. Rev. Lett.*, **59**(1987), 2911.
[111] C. Maneveau and K. R. Sreenivasan, *Phys. Rev. Lett.*, **59**(1987), 1424.
[112] B. B. Mandelbrot, In *Random fluctuations and pattern growth*, H. E. Stanley and N. Ostrowsky (eds) (Kluwer Academic, Boston, 1988).
[113] B. B. Mandelbrot, *J. Fluid Mech.*, **62**(1974), 331.
[114] M. Takayasu and H. Takayasu, *Phys. Rev.*, **A39**(1989), 4345.
[115] T. Tel and T. Vicsek, *J. Phys.*, **A20**(1987), L835.
[116] J. W. Ward, R. L. Kubena and M. W. Utlant, *J. ac. Sci. Technol.*, **36**(1988), 2090.

Further reading

Introduction of fractals

Mandelbrot [1].
[117] J. Feder, *Fractals* (Plenum, New York, 1988).
[118] T. Vicsek, *Fractal growth phenomena* (World Scientific, Singapore, 1989).

Mathematical aspects

Falconer [105].
[119] H.-O. Peitgen and D. Saup (eds), *The science of fractal images* (Springer, Berlin, 1988).
[120] M. Barnsley, *Fractals everywhere* (Academic Press, San Diego, 1988).

Physical sciences

Liu [74].
Pietronero and Tosatti [108].
[121] R. Pynn and A. Skjeltorp (eds), *Scaling phenomena in disordered systems* (NATO ASI series B133, Plenum, New York, 1985).
[122] H. E. Stanley and N. Ostrowski (eds), *On growth and forms, a modern view* (Kluwer Academic, Boston, 1985).
[123] N. Boccara and M. Daoud (eds), *Physics of finely divided matter* (Springer, Berlin, 1985).
[124] H. E. Stanley and N. Ostrowsky (eds), *Random fluctuations and pattern growth* (Kluwer Academic, Boston, 1988).

Chaos

[125] H. G. Schuster, *Deterministic chaos* (Physik Verlag, Weinheim, 1984).
[126] H. O. Peitgen and P. H. Richter, *The beauty of fractals* (Springer, Berlin, 1986).
[127] P. Fisher and W. R. Smith (eds), *Chaos, fractals and dynamics* (Marcel Dekker, New York, 1985).
[128] J. Gleick, *Chaos, making a new science* (Viking Penguin, New York, 1987).

Percolation

Stauffer [60].
[129] H. Kesten, *Percolation theory for mathematicians* (Birkhäuser, Boston, 1982).

Aggregation

Jullian and Botet [17].
[130] H. Herrman, *Phys. Rep.*, **136**(1986), 153.
[131] F. Family and D. P. Landau (eds), *Kinetics of aggregation and gelation* (North-Holland, Amsterdam, 1984).

Index